HOW SAFE IS FOOD IN YOUR KITCHEN?

Other Books by Beatrice Trum Hunter

THE NATURAL FOODS COOKBOOK
GARDENING WITHOUT POISONS
CONSUMER BEWARE! YOUR FOOD AND WHAT'S BEEN DONE TO IT
THE NATURAL FOODS PRIMER: HELP FOR THE BEWILDERED BEGINNER
WHOLE-GRAIN BAKING SAMPLER
FOOD ADDITIVES AND YOUR HEALTH
YOGURT, KEFIR AND OTHER MILK CULTURES
FERMENTED FOODS AND BEVERAGES: AN OLD TRADITION
BEATRICE TRUM HUNTER'S FAVORITE NATURAL FOODS
THE MIRAGE OF SAFETY
THE GREAT NUTRITION ROBBERY

HOW SAFE IS FOOD IN YOUR KITCHEN?

Beatrice Trum Hunter

CHARLES SCRIBNER'S SONS
NEW YORK

To Harold

Copyright © 1981 Beatrice Trum Hunter

Library of Congress Cataloging in Publication Data
Hunter, Beatrice Trum.
How safe is food in your kitchen?
Bibliography: p. 80
Includes index.
1. Food contamination. 2. Kitchen utensils.
I. Title.
TX535.H78   363.1'9   80-25635
ISBN 0-684-16752-2

Contents

Preface
vii

PART I
KITCHEN UTENSILS
1

1 *Pots and Pans*   3
2 *Kitchenware*   33

PART II
SOME COOKING DEVICES
49

3 *Microwave Ovens*   51
4 *Charcoal Grills*   57
5 *Food-smoking Chambers*   60

PART III
DRINKING WATER
63

6 *Water from Your Tap*   65
7 *Household Water Softeners*   69
8 *Household Water Filters, Purifiers, and Distillers*   72

Principal Sources
80
Index
85

Preface

This handbook demonstrates that in your kitchen many metals, chemicals, plastics, and other substances in contact with food and water can migrate and interact. To what extent are these interactions dangerous? The handbook is intended to be a practical guide, answering many commonly asked questions. What pots and pans are safe? Should I discard aluminum utensils? Is nonstick cookware hazardous? What's wrong with charcoal broiling or smoked food? Are microwave ovens safe? Is plastic food wrap dangerous?

These questions deserve answers based on reliable information, but no governmental agency, nor any single reference

source, deals comprehensively with the subject of food contacts. Scattered articles, mainly written in technical lan-
guage, are available but are neither readily accessible nor easily understood.

How Safe Is Food in Your Kitchen? is intended to fill this
void. Its purpose is to present reliable information and to separate facts from fallacies. Enlightened consumers are
equipped to make intelligent choices in the marketplace as
well as to control the safety of their food and water.

PART I

KITCHEN UTENSILS

1

Pots and Pans

Any defects in food may have an effect on health which may not be apparent for some time and for which it is often difficult to account. Perhaps the damage may take a long time to repair, as for instance, in certain forms of metal poisoning. Food is the one commodity the quality of which must, in the interests of the consumer, be most closely scrutinized and most rigidly controlled.
—DR. G. W. MONIER-WILLIAMS
British Ministry of Health, 1935

Ideally, your pots and pans should be made of inert substances that do not peel, chip, dissolve, craze, crack, vaporize, or migrate from the utensil into the food, and should be good conductors of heat that cook food uniformly throughout. Unfortunately, no one type of cookware has yet been manufactured that achieves all these goals. Some types of pots and pans are better than others for specific purposes, but all have certain characteristic shortcomings as well as desirable features.

Fortunately, you have many choices of pots and pans, depending on your style of cooking. What appeals to you may not necessarily appeal to others. If you are worried about using certain types of cookware—whether the concern is valid or not—you can easily avoid cookware you consider of questionable safety.

An alphabetical listing of the various types of cookware follows, with an indication of the advantages and shortcomings of each.

AGATEWARE: *See* Enamelware

ALUMINUMWARE

Aluminum utensils, first introduced in America in 1902, became popular because aluminum conducts heat rapidly and evenly. Aluminum pots and pans are comparatively inexpensive, handle easily because they are light in weight, and are attractive.

For many years, the safety of aluminum cookware has been debated. The FDA, for example, has claimed that there is no evidence that aluminum cookware, containers, or foil materials are unsafe in food contacts. The agency has claimed that "salesmen of cookware not made of aluminum have tried to convince consumers that cooking with aluminum utensils is injurious to health and should be discarded in favor of the salesmen's own wares. These salesmen point to the greyish looking substance that gradually accumulates on the surface of aluminum and say that this could be harmful to foods cooked in the aluminumware. This greyish substance is harmless. Some aluminum compounds are even used as food ingredients, for instance alum in pickling and these are generally recognized by scientists as safe."

The FDA's statement is a mixture of fact and foolishness. True, some sellers of nonaluminum cookware may use high-pressure sales pitches to promote their products (usually of stainless steel) and scare tactics to denigrate aluminumware.

True, too, a grayish residue gradually accumulates on the surface of aluminumware after certain types of foods or liquids have been cooked in it. But it is foolish to compare this residue, dissolved from cookware, to aluminum compounds used for specific purposes as food additives. Different compounds from the same base may act quite differently with regard to human health. For example, table salt (sodium chloride) and laundry bleach (sodium hypochlorite) are both sodium compounds; while salt is added to food, laundry bleach is not.

In claiming "no evidence" of harmfulness to human health from aluminum cookware, the FDA chose to ignore a sizable number of reports, many appearing in respectable medical journals. Granted, most of them are anecdotal, old, inadequate, inconclusive, and lack "controls" of scientific studies, but such evidence, with all its limitations, should not be totally ignored.

What are the known facts about aluminumware? Usually aluminum is only very slightly dissolved by solid or liquid food. But three types of food dissolve significantly more aluminum when the utensil is used for cooking and storing:

- acidic foods, such as fruits, apple butter, cranberry sauce, rhubarb, and tomatoes, as well as liquids such as fruit juice or vinegar
- alkaline foods or liquids such as hard water or detergent which interact with the metal and form a coating which stains the utensil; this gray powdery coating, a harmless oxide or "rust," can be removed through scouring
- salty water (brine) or salty food pit aluminumware

Pitting also occurs if the water contains even minute quantities of copper or other metals. Trace amounts of arsenic and fluorides, impurities frequently present in cooking-utensil grade of aluminum, also may be dissolved along with the aluminum from the utensil. These metals may be more hazardous than the aluminum.

Many foods contain some aluminum, although the concentrations are low, especially in foods of animal origin. Concentrations range from half a part per million in raw bananas and apples to 10 parts per million in baked potatoes, 15 parts per million in wheat bran flakes, and 120 parts per million in fudge cake. The highest value found—260 parts per million—was in waffles made with an aluminum-containing baking powder. Examination of all the determinations of the aluminum content of foods, both raw and cooked, led to the conclusion that the quantities of aluminum in foods, after being cooked in aluminum utensils, are well within the range of concentrations found in raw products. Actually, many foods cooked in aluminum pots were found to contain *less* aluminum than they did in the raw state. Two explanations are possible. Dust or soil from the earth is the principle source of aluminum in most foodstuffs. This source adds aluminum to raw fruits and vegetables. Potato parings, for example, have much more aluminum than the interior of the potato—as much as 140 times. And vegetables gathered from the home garden may carry more aluminum to the table than vegetables that have been cleaned and processed in canneries and freezing plants, even though they may have been processed on aluminum machinery.

In the 1930s, the Mellon Institute of Industrial Research conducted studies of foods cooked in aluminumware. Food samples included apple butter, creamed cabbage cooked with baking soda, percolated coffee, apricots, rhubarb, cranberry sauce, tomatoes, oatmeal, lemon pie filling, creamed chicken, and boiled ham. Some aluminum was dissolved in *all* of the food samples. A metal spoon, used for stirring, set up a galvanic current and increased the amount of aluminum that went into solution and combined with the food.

Although the Mellon Institute's tests demonstrated that aluminum dissolved from cookware and entered the food, the purpose of the research was not to explore what happens in the human body when aluminum is ingested. This question is the crux of the controversy.

The public's concern about the use of aluminum cookware continued. In the 1950s, the Kettering Laboratory of the Department of Preventive Medicine and Industrial Health at the University of Cincinnati undertook to assemble existing data about the relationship of aluminum and human health. Staff members reviewed some 1,500 books and articles written by investigators in more than 20 countries. The final report concluded that "there is no reason for concern on the part of the public or of the producer and distributor of aluminum products, about the hazards to human health derived from well-established and current uses of such products. Nor need there be concern over the more extended uses which would seem to be in the offing."

Despite the reassurances of two prestigious institutions, Mellon and Kettering, wariness about using aluminumware persisted with segments of the public. What are the toxic effects, if any, of aluminumware on humans? The answer remains equivocal. Statements alleging toxicity from small amounts of ingested aluminum are based largely on clinical evidence which is difficult to assess. Cause and effect have not been established.

Individual idiosyncratic reactions to various substances, including metal, are reported. Possibly some individuals are extremely sensitive to aluminum. In the literature of metal hypersensitivity, however, metals such as copper and nickel are frequently noted, while aluminum is not.

Internationally respected experts in trace metals concluded from their work that aluminumware posed no health hazards. The late Dr. G. W. Monier-Williams, E. J. Underwood, and Henry A. Schroeder, M.D., all expressed the belief that cooking in aluminum pots and pans was a safe practice. Schroeder considered aluminum as "probably inert" in the human body. With wry humor, he classified this metal in his "glossary of worries" as one that created "no problem in pots . . . a problem in beer cans on roadsides."

Nevertheless, the traditional belief that aluminum is not toxic has been challenged anew. The subject of aluminum cookware is related to a larger issue. Residents in Albany Village, California, living close to a factory that released fine aluminum powder into the air, took legal action. They claimed that the aluminum dust which had been collecting as a silver film on cars, furniture, and household plants, was also inflicting human health problems. A study of the aluminum accumulation in the bodies of 46 Albany Village residents showed that 84 percent had higher-than-normal levels of aluminum and 30 percent had significantly high concentrations. Children were more apt to accumulate significantly high levels.

Commenting on the Albany Village study, Dr. Carl C. Pfeiffer, director of the Brain Bio Center in Princeton, New Jersey, said that inhaled or ingested aluminum accumulates in the brain, and in time may cause memory loss and brain deterioration. He charged that high levels of aluminum in the brain have been associated with Alzheimer's disease, the most frequent cause of senile dementia at an early age.

Pfeiffer's charges were refuted by Dr. Lorne Cassidy, the chief physician for the aluminum company, who reported that aluminum had not been found to *cause* Alzheimer's disease, but rather that the disease itself might alter the body's chemistry and allow aluminum to accumulate in the brain.

The environmental health officer in the Albany Village area reviewed the scientific literature on aluminum toxicity and announced that it was "contradictory and inconclusive." And so the debate continues to rage.

What is known and what is suspected? The low levels of aluminum that may be ingested combine with phosphorus in the human intestine and form an insoluble compound, aluminum phosphate. This compound is excreted in the feces, and confirms Schroeder's statement that aluminum is "probably inert." The process itself, however, may not be innocuous but may create phosphorus or calcium deficiencies. When aluminum binds phosphorus as aluminum phosphate the remaining phosphorus level may be inadequate for proper functioning of the sympathetic nervous system. Also, the phosphorus may no longer be available for calcium absorption and utilization. There has been a suggestion that aluminum, by depriving the body of sufficient calcium, may be responsible, indirectly, for many allergic symptoms. Allergic patients, exposed to aluminum, were reported improved with calcium supplementation. Aluminum compounds that may deplete the body of phosphorus and calcium are used extensively in consumer goods (deodorants), over-the-counter drugs (antacids), baking ingredients (aluminum baking powders), many food additives (anticaking agents, dispersing agents, binding agents, emulsifying agents, defoaming agents, modifying agents, and certain food colors), metal cans (containing soft drinks and beer), and toiletries (toothpaste), in addition to pots and pans. Also to be considered are the millions of pounds of aluminum powder added to municipal water supplies to speed the precipitation of sediments.

It should be apparent that any examination of the health effects, if any, resulting from the use of aluminum cookware, needs to be explored in relationship to *all* sources of environmental exposure to aluminum. There is general consensus among those who have studied aluminum and human health that there is no convincing evidence of aluminum toxicity from cookware in the amounts likely to be consumed by the average person. The caveat may need to be added that in some rare cases, it is conceivable that individuals may be highly sensitive to specific metals, including aluminum, and need to avoid *all* sources of the metal.

If you decide to purchase aluminumware, remember that the thicker the utensil, the better the heat will be held and distributed. Cast aluminum is heavy. If you wish to keep the aluminum bright and minimize interactions with food, don't cook or store acidic, alkaline, or salty food or liquid in aluminum utensils. The cookware will stay shiny and new looking if you scour it whenever necessary.

If you have strong gut feelings that aluminumware should not be used, exercise your prerogative and avoid it. If you suspect that you may be sensitive to aluminum, by all means avoid it. You may choose utensils of many other compositions.

BAKED ENAMEL COOKWARE: *See* **Porcelain-enamel Coated Cookware**

CAST IRONWARE: *See* **Ironware**

CERAMIC COATED COOKWARE: *See* **Porcelain-enamel Coated Cookware**

CLAY EARTHENWARE
If you enjoy cooking one-pot casserole meals with gentle, long oven cooking, clay earthenware will appeal to you. Flavors of stew components blend together in this style of cookery, and inexpensive meats that require long, moist cooking do well in such cookware. An example is the crockpot.

Clay earthenware may be either unglazed or glazed. The popular terra-cotta-type casserole is similar to its predecessor in the ancient Roman world. You'll need to handle such cookware carefully, for it is fragile and may crack or break when subjected to sudden temperature change. Being porous, such cookware tends to retain traces of food, resulting in microbial contamination. But unglazed earthenware is free of lead.

Glazed clay earthenware is coated with a compound glass that seals its surface. You'll find this type easier to keep clean, and more durable than unglazed clay. The glaze also keeps food from being absorbed, and microorganisms from being entrapped on the container's surface.

Glazes for clay earthenware contain both lead and cadmium, as well as other toxic metals. Their presence is not hazardous unless they leach from the glaze and get into food or beverages. The critical factor is how resistant the glaze is to attack by acids. Some, but not all, glazes contain toxic acid-soluble metal compounds that can be leached in varying amounts by acid-containing foods and liquids such as sauerkraut, tomatoes, tomato products, foods containing vinegar, fruit juices, soft drinks, wines, and cider. The toxic metals leach out of the glaze to a greater degree if the food or beverage is warm or hot rather than if it is cold. The longer the storage time, the greater the amount of leaching.

The leaching of lead and cadmium from glazed earthenware pots, as well as from dishes, is a problem well recognized by authorities. The FDA monitors such utensils and frequently seizes shipments of clay products, especially imported ones, containing excessive levels of these toxic metals. Domestic earthenware seizures are less common, for the industry is knowledgeable and skillful in the formulation and application of glazes.

Homemade pottery, however, does not have the same safety record. At times, glazes are poor and products are dangerous when used as food and beverage containers.

Sometimes, decals used to decorate earthenware are applied over the basic glaze and bonded by firing. Decals, too, may contain lead and cadmium.

Since clay vessels are in worldwide use, especially in developing countries, lead and cadmium poisoning from earthenware vessels has been a widespread problem. In 1976, the World Health Organization established tolerance levels for toxic metals in earthenware, applicable on an international scale. The following year, the Department of Ceramics, College of Engineering, Rutgers University, was designated as secretariat to a group studying and monitoring the problem.

If you use unglazed casseroles, scrub them scrupulously after use to eliminate food residues. To prevent an accumulation of rancid fat, fill the pot with a solution of baking soda and water, and bake for an hour at 400° F. If you plan to buy glazed earthenware, choose domestic products made by reputable manufacturers, not cheap imports with questionable glazes. Pass up any earthenware that is decorated with decals. If the earthenware glaze shows any signs of cracks or crazes after it has been in use, discontinue cooking with it. Don't store foods and beverages in earthenware. Play it safe, and limit the use of homemade pottery to decoration.

COPPERWARE

Aesthetically you may enjoy the cheery appearance of bright copper utensils in your kitchen, and many pots made of poor heat conductors have copper-clad bottoms, for copper is one of the best heat conductors. But if you choose a copper pot as a cooking utensil, make certain that the vessel is lined so that the copper itself is not in contact with food.

Copper poisoning from food contact is neither new nor rare. In 1785, Dr. Thomas Percival of Manchester, England, reported to the Royal College of Physicians in London that the careless use of copper pots "puts us in daily hazard, not only of the loss of health but even of life itself." The hazard was well recognized by the early 1800s, when it was common practice to use copper compounds to enhance the appearance of some foods. Copper arsenite was added to color tea leaves and to make puddings green. Candies with attractive rainbow hues, commonly sold in London sweetshops to children, were colored with poisonous copper salts. Pickles, colored bright green with copper sulfate, caused fatalities. The practice of adding copper sulfate continued, even into the twentieth century, to bestow an artificial greenness to cooked peas, even though copper poisoning from food contacts was an acknowledged fact.

As recently as 1977, a mail-order catalog for gourmet cooks listed an unlined copper preserving pan, accompanied by this blurb:

> *Larousse Gastronomique* says firmly that you cannot make jams and jellies without an untinned copper or aluminum pan, the reason being that the heat MUST be conducted evenly, and the pan has to be wide so that evaporation can take place in the first stage of cooking and impurities can be easily skimmed off. The "jelling" process occurs in the second stage of cooking, quickly or slowly, depending on the type of fruit. Our magnificent heavy unlined French copper preserving pan holds twelve quarts.

Twelve quarts of jam may become twelve quarts of poison. The unlined copper preserving pan, in contact with acidic fruits cooked for a long time, can leach toxic levels of copper from the vessel and poison the preserve.

Gourmet food-equipment catalogs list unlined copper measuring cups and unlined copper bowls for beating egg whites stiffly. Since these accessories have only brief contact with food, and are not intended for cooking use, they probably are not hazardous. But if you were to use them for other purposes, such as for storage, they might be hazardous.

As recently as 1975, two incidents of mass copper poisoning were reported by the Center for Disease Control (CDC). The first episode was an outbreak of copper poisoning in a nursery Bible class in Mesa, Arizona. Eight young children, aged one to four, became ill shortly after drinking an orange-flavored beverage stored in a brass pot overnight. Brass contains copper. Analysis of the drink showed a high copper level. CDC issued a warning against the preparation or storage of any beverage in brass containers.

The second incident reported by CDC involved a dozen guests at a wedding reception in Montgomery County, Pennsylvania, who suffered acute copper poisoning after drinking punch or whiskey sours from fountain-type containers with copper undercoating.

Although a small amount of copper is harmless, and indeed essential to human life, high levels are toxic and may deplete the body of another essential element, zinc. Individuals with Wilson's disease, a genetic disorder, may need to avoid all contacts with copper, including the minute amounts that may leach even from apparently sound copper cookware.

Buy only *lined* copperware for cooking purposes. Examine the lining periodically to make sure it is in good condition. If the lining becomes worn, have it relined or discontinue using the utensil with food. Keep your copper burnished by rubbing it with a soft cloth dipped in a solution of salt and vinegar.

CROCKPOT: See **Clay Earthenware**

EARTHENWARE: See **Clay Earthenware**

ENAMELWARE
Good-quality, sound enamelware is serviceable for pots and pans. If you handle them carefully, such utensils will remain sound, will not chip easily, nor leach components. Intact, they are inert. To make the enamel white and

opaque, the manufacturer uses tin oxide, an expensive and satisfactory compound that makes enamelware safe.

However, if you buy very inexpensive enamelware you are indulging in a foolish economy. The enamel on cheap utensils is much softer, not acid resistant, and will chip readily. These enamels are composed of mixtures low in silica, which allows them to fuse easily and inexpensively, and be fired at a low temperature. They are made with antimony oxide, which is cheaper than tin oxide, but hazardous. If the enamelware chips, the antimony oxide leaches into the food and goes into solution as a toxic compound, tartar emetic. There have been numerous instances of people poisoned at social gatherings by drinking acidic beverages such as lemonade, prepared and stored in soft enamel vessels.

Good-quality, sound enamelware, according to tests conducted by Professor Franco Marzulli Modeno, an Italian microbiologist, proved highly resistant to bacterial contamination and fungal growth in food. Modeno tested cooking utensils of different composition and their effects on food. Enamelware showed the lowest level of bacterial and fungal growth. Aluminumware had fifty-four times as much contamination; nonstickware seventy-six times as much; and anodized aluminumware, eighty-two times as much.

Invest in good-quality utensils and treat them carefully. In recent years, manufacturers have tried to make enamelware utensils more durable by making them more resistant to chips and cracks. If a chip or crack develops on the outside of the utensil, the blemish may not be aesthetically pleasing, but it is not hazardous. If the chip or crack is in the interior where it will be in contact with food, discontinue using the utensil. To be on the safe side, don't use enamelware to store acidic food, even if the enamel is sound. You may not want to choose enamelware utensils if you feel that you have to exercise special care to keep them from being marred. Years ago, enamelware was called agateware or graniteware, and these terms are still applied occasionally.

FLUOROCARBON-COATED NONSTICK COOKWARE: *See* Nonstick Cookware

GLASS COOKWARE

You may like glass cookware because it is sanitary. Being inert, glass does not leach anything into food. It does not peel or chip. Food residues are easily removed from glass. You may enjoy a glass pot for brewing coffee or tea since the pot does not impart any off-flavor to the beverage. If you happen to suffer reactions to metals, even at exceedingly low levels, you may need to use glass cookware.

But glass cookware has some disadvantages. You need to handle glass carefully to prevent cracks or breakage. You need to use gentle heat, without any sudden temperature change. Glass, a poor conductor, heats slowly and does not distribute heat evenly. A glass skillet will not brown food. Cooking foods on top of the stove in glass cookware exposes them to light, and foods that are good sources of riboflavin (vitamin B_2), including many green vegetables, grains, and stewed fruits, may lose some of this valuable nutrient which is light sensitive.

Check the label and look for the word "ovenproof." If the glass cookware is only "flameproof," you can only use it for top-of-stove cooking. To minimize breakage on top-of-stove cooking, use a metal grid (sometimes called a flame tamer) between the heat element and the utensil. Always begin cooking with low heat. In using glass cookware for baking, adjust the time and temperature of the oven. If bread bakes for an hour in metal pans, it may need only fifty minutes in glass. Set the oven temperature from 25° to 50° F. lower.

Consider inert opaque ceramic glass cookware as an alternative to glass cookware. This attractive cookware is less fragile and does not expose food to light. It is practical, since you can cook, serve, and store food in it. But, like glass, opaque ceramic glass heats slowly and distributes heat unevenly.

Clean both glass and opaque ceramic glass with care, for harsh abrasives will scratch their surfaces.

GRANITEWARE: *See* **Enamelware**

IRONWARE

Whether your cooking style entails gentle sautéing or quick, steady heat, you'll enjoy using cast ironware. The utensils are durable and, with care, are a lifetime investment. Ironware also yields nutritional benefits, a unique claim for cookware.

Iron pots have been used since prehistoric times, and they are still used widely in most parts of the world. The Chinese may have been the first to use cast-iron pots. They produced cast-iron bowls for cooking and eating rice nearly 3,000 years ago. In Europe, during the Middle Ages, cast-iron utensils were deemed so valuable that they were listed along with gold, jewels, and other treasures of royalty. One of the first industries to be organized when the North American continent was settled by Europeans was an iron foundry for the manufacture of cast-iron products including cooking utensils. Unfortunately, when stainless steel and other nonferrous cookware became popular in America, many iron pots were discarded, because it is true that they are heavy to handle and require some care to prevent rusting.

One of my happiest childhood memories was the savory pot roast my mother prepared in a well-tempered cast-iron Dutch oven. But she, like many homemakers of the time, wearied of the pot's weight and what she felt was the nuisance of having to treat it after each use to prevent rust. In tune with the time, she threw out the pot and replaced it with a nonferrous one. For a long time, our family suffered silently. The pot roasts were fair, but not comparable to those we had come to enjoy. We ate politely but unenthusiastically. At last, on one Christmas morning my mother was presented with another cast-iron Dutch oven, purchased jointly by all members of the family. Once again, we were able to enjoy those delectable pot roasts, which could only be cooked to perfection in ironware.

The replacement of ironware by non-ferrous cookware resulted in nutritional as well as flavor losses. Ironware releases some iron into food, a *rare* instance when metal leached from a utensil into a food is regarded *favorably*. Although the form of iron dissolved from pots may not be easily absorbed and utilized, iron pots do interact with foods, release some iron into them, and contribute to the total level of dietary iron. For example, it is estimated that eggs cooked in an iron skillet triple their iron content. In 1971, the late Dr. Carl V. Moore compared the iron content of spaghetti sauce cooked in iron Dutch ovens with that of spaghetti sauce cooked in glass. The sauce cooked in ironware showed nearly thirty times as much iron as the sauce cooked in glassware.

The incidence of iron-deficiency anemia, reported to be widespread in America, especially among young children and women of childbearing age, might well decrease if more people used ironware for cooking. However, persons suffering from hemochromatosis accumulate too much iron and need to avoid food contacts with ironware.

Buy the best quality ironware. Some cast ironware is "preseasoned" by the manufacturer. If it isn't, follow the manufacturer's printed directions for "seasoning" the utensil. It should not be washed with strong detergents or scoured. Instead, use hot soapy water, rinse, and wipe it dry immediately. Make certain that it is thoroughly dried. Never store a cast-iron utensil with the cover on, as this may cause sweating and rust damage. Store iron utensils in a dry place.

The inside of the washed and dried utensil should be treated frequently with a coating of unsalted fat or cooking oil. Before using it again, simply wipe the utensil with a dry cloth or paper towel. If you reserve one ironware skillet solely for egg cookery, wipe the pan thoroughly but do not wash it.

If necessary, you can reseason a cast-iron utensil. Scour the utensil thoroughly, wash it in soap and water, dry thoroughly, and then coat the inside surface with unsalted fat or cooking oil. Place the utensil in a moderately heated oven for about two hours, remove, and wipe off any excess fat or oil.

Since you can turn off the heat element under ironware and the utensil will retain heat longer than other types of cookware, ironware can be an energy saver. You can also use an all-iron utensil, even a skillet, in the oven as well as on top of the stove.

NONSTICK COOKWARE

You may enjoy the ease of cleaning nonstick cookware, which has become one of the most popular types of cookware in American homes since the mid-1960s. The main advantages were stressed in the early sales pitch: "no-stick cooking; no-scour cleaning." In addition, if you are on a low-fat diet, it is possible to fry foods without fat or oil.

Despite these benefits, the use of nonstick cookware has been controversial since it was introduced. Let's separate facts from fallacies.

In 1938, Du Pont researchers accidentally discovered an extremely durable resin, tetrafluorethylene (TFE), given the trademark name Teflon. The resin, extremely resistant to corrosive chemicals, found many industrial applications. Its nonstick quality made it useful as a coating for cookie sheets and pans used by bakers and candymakers, and for milk handling equipment used by dairymen. Soon the Du Pont resin was applied by cookware manufacturers to frying pans, and these nonstick pans became popular in many American homes.

The first doubts raised about the safety of Teflon were based on experiences of industrial workers exposed to TFE fumes. TFE resin, heated to a temperature of about 200° C. (737.6° F.), releases highly toxic fumes. Apparently this effect triggered wild rumors. By the mid-1950s, an anonymous account, widely circulated, concerned a machinist who died after smoking a cigarette contaminated with TFE resin. Variations of this account persisted. Du Pont employed a public relations man who worked nearly full-time answering letters from worried consumers. *Anatomy of a Rumor*, written by the director of Du Pont's Haskell Laboratory for Toxicology and Industrial Medicine, is an account of the scare story. Du Pont officials doubt that the rumor will ever be laid to rest. The story not only persisted, but spread. In France, the resin-coated pan was subjected to the scrutiny of three major laboratories and to, according to *Anatomy of a Rumor*, "what may well have been the most extensive scientific scrutiny of any kitchen utensil in history," as the result of the uproar. All three laboratories concluded that the pans created no health hazard.

The FDA also felt the impact of public anxiety, and the agency was deluged by inquiries. At one time, the FDA considered matters such as the safety of cookware outside its jurisdiction. But no federal government agency was responsible for determining the safety of materials used for cookware and their coatings. In the early 1960s the FDA reversed its position. Although the agency decided against issuing any formal approval of resin-coated nonstick cookware, FDA pharmacologists stated that there were "no foreseeable hazards through use of these [resin-coated] frying pans in normal cooking."

Despite assurances from Du Pont, the FDA, and three

French testing laboratories, public worry continued. In 1971, in response to an unrelenting stream of inquiries, the FDA issued an information sheet. Safety data had been submitted by the manufacturer for resins used in commercial food preparation. The testing conditions were reasonably similar to home use of the frying pan. As could be expected, the resin would decompose if dry pans, without food, were heated at high temperature. However, the temperature at which decomposition occurs is well above the smoke point of cooking fats. During cooking, significant decomposition would not occur. The toxicity of fumes given off by a dry heated pan was somewhat less than that of fumes given off by ordinary cooking oils.

The FDA noted that some transfer from cookware to food occurs even with such durable utensils as stainless steel cookware. The agency tested various types of pans to learn whether continued use of resin-coated cookware increases the possibility of food contamination. Four types of pans were used: a newly resin-coated pan; an aged pan (heated at 250° F. for 150 hours to age it); a pan that had had two and a half years of domestic use; and a control pan of uncoated aluminum. Samples of hamburger were fried in all four pans, and the cooked meat was analyzed for possible fluoride contamination. (The presence of fluoride was judged as an indicator of tetrafluorethylene released from the resin, or from the aluminum, which contains trace fluoride as an impurity.) Results showed that the hamburger cooked in the new resin-coated pan contained approximately the same quantity of fluoride as the hamburger cooked in the uncoated aluminum one. A slight increase of fluoride was found in the meat cooked in the heat-aged and in the old pans, but the amount of fluoride was well within a safe limit. The FDA scientists concluded that resin-coated utensils are safe for conventional kitchen use.

Is resin-coated cookware safe? The well recognized hazards

of TFE fumes in industrial situations are not comparable to home use of resin-coated cookware. Even under extreme conditions, when cookware is inadvertently overheated, the resin fumes emitted would be slight, since the amount of resin used to coat the utensils is minimal. The fumes would be released into a large volume of kitchen air, and the exposure time would be brief. Such conditions are not similar to industrial ones, where fluorocarbon resins decompose at very high temperature.

Nevertheless, if you use nonstick cookware in your kitchen, you should realize that food cooked in such utensils needs to be tended and not overheated. In Germany, the government requires that resin-coated utensils bear the warning "Dangerous when overheated."

If you are sensitive to fluorides and need to avoid them, don't use resin-coated nonstickware or aluminumware, for both release small quantities of fluoride.

Buy good quality nonstick utensils from a reputable manufacturer. As the resin coating began to have far wider application, from the frying pan to many types of cookware, newer methods of production made possible more durable finishes. The finish may now include nonstick high-temperature resin, silicone or fluorocarbon resin, or combinations of these materials.

Nontoxic plastic resins, capable of withstanding cooking or baking temperatures, are made of polyamide-imide or polyphenylene sulfide. Occasionally these resins contain some fluorocarbon resins as well to improve the nonstick quality. The characteristics of such cookware are its nonstick property, ease of cleaning, safety in the dishwasher, chemical inertness, and color.

Silicones are nontoxic synthetic resins applied to specially prepared metal surfaces and joined to the metal by bonding. Silicone finishes are applied to the interior and exterior metal surfaces of bakeware. The characteristics are its nonstick property, ease of cleaning, excellent gloss, and color.

Fluorocarbon resins are solid nontoxic plastic polymers that have an inherent nonstick property. These materials are generally applied in one, two, or three coatings. Some trade names identifying nonstick fluorocarbon finishes are "Teflon," "SilverStone" (both are registered trademarks of E. I. Du Pont de Nemours Co.), "Fluon," "Debron," "Xylan," and "Rock-Bottom." Du Pont's SilverStone, which has a pebbly texture, is applied in triple coatings. Fuse bonding, that is, fusing several layers at 800° F., can be done only with at least an eight-gauge thickness on skillets; ten, on saucepans. (The gauge of aluminum is determined by the number of sheets required for an inch of thickness. Thus, ten-gauge metal is *thinner* than eight-gauge, which requires only eight sheets to make an inch.) Ask about the gauge and number of coatings used on cookware before you buy it.

Many manufacturers of nonstick utensils recommend that frying pans, Dutch ovens, and bakeware be seasoned with cooking oil before they are used for the first time. Cooking oil should be wiped on the nonstick surface with a paper towel. An exception is the tubed angel food or sponge cake pan, which should not be oiled or greased if used for such cakes because the batter needs to cling to the sides of the pan during baking time. For baking cakes that contain fruit or other sweetening agents, some greasing is usually necessary to ensure complete release of baked goods from the surface of the utensil.

The newer coatings do not scratch as easily as older or thinner ones, so you can use metal implements with them. But there is a trade-off. Although the newer coatings are more durable, their nonstick quality is not as good, and the coatings discolor more readily.

If you own older nonstick utensils, use wooden or soft plastic implements for stirring or scraping food. If the coating becomes scratched, peels, or chips, discontinue using the utensil with food.

If you hesitate to use nonstick cookware, but enjoy the nonstick characteristic, spread a small quantity of liquid soy lecithin into any type of utensil to coat its surface before you use it. This way, you don't have to use nonstick cookware, or commercial sprays sold for this purpose.

Or invest in a soapstone griddle, which does not require fat or oil. These griddles are available from mail-order companies, but their supply tends to be irregular. Two sources are the New Hampton General Store, Box 71, Hampton, N.J. 08827, and the Vermont Country Store, Weston, Vt. 05161.

OPAQUE CERAMIC GLASS COOKWARE: See Glass Cookware

PLASTIC COOKING BAGS

Are you among those who enjoy the use of many foods now packaged in plastic cooking bags and boil-in-bag pouches? These convenient containers eliminate messy pots and pans.

At first glance, plastic cooking bags seem like a blessing. Meats roasted in such bags baste themselves. Residues of food fats and juices don't have to be removed from pots. Commercially frozen cut vegetables, sealed by the processor in the plastic pouch, are merely lowered into boiling water for cooking.

Formerly, it was thought that plastics used in such bags and pouches were inert, but accumulating evidence shows that they are not. Under certain conditions, plasticizers (chemicals used to keep the plastics soft) can be leached from plastics. How safe is food when plasticizers enter them? Since the release of plasticizers may increase with high temperature, what happens when the plastic is in contact with hot meat fat and juices?

To explore these questions, a researcher checked on plastic cooking bags and pouches available in supermarkets. The manufacturers of two products listed the trade names of the plastic films used, while a third did not, but warned, "Do not use over 425° F." There was no way to know which plasticizers may have been used in any of these products.

The warning against overheating the plastic is justified. Nylon films, degraded by heating to 581° F. (305° C.) mainly produce carbon dioxide, water, and ammonia. But small quantities of various amines are formed, too, including highly toxic hexamethyleneamine, slightly toxic *n*-hexylamine, and methylamine (toxicity unknown).

In September 1970, the Reynolds Metals Company began marketing cooking bags with nylon film, cuprous iodide, and cuprous bromide. Although each of these food additives had been approved by the FDA individually, their combined use had not. At the time, the FDA considered plastic cooking bags as utensils, not subject to food-additive laws. Two years later, the agency reversed its position, and, in February 1972, asked Reynolds to stop shipping its cooking bags temporarily on the grounds of a technical violation. By then, cooking in plastics was no longer *assumed* safe; additional problems involved fires and burns.

The New York State Department of Health recorded nearly a hundred accidents involving oven fires and burns from hot juices splattered from burst cooking bags. As a result, the FDA required printed instructions on the package. They recommended using a pan large enough to hold the juices in the event of the bag's breaking, and keeping the oven door closed in the event of a fire.

Since the bags were first introduced, many specialty pouches have been developed for certain foods. A boil-in-bag rice is packaged in a film coextruded with polyethylene layers of different densities. The bag, perforated with numerous holes, allows the boiling water to flow through the film into rice treated to cook in a very short time. This type of pouch is called a "colander bag." Its convenience comes at a high price.

Another specialty boil-in-bag pouch product is for tortillas, already fried and ready to be boiled, filled, and baked. The purported convenience is that it spares the homemaker the necessity of frying the tortillas to soften them before further preparation. The tortillas are wrapped in plastic film coextruded with mylar (polyethylene terephthalate) and polyethylene layers to protect the product against oxygen, moisture, and rancidity.

Many boil-in-bag pouches are made of polybutene, a plastic film that withstands temperatures above boiling. Others are made from coextruded layers of low-density polyester film and medium-density polyester web.

What price do you pay for such convenience? According to Esther Peterson, consumer advocate, "In some cases, the convenience is illusory. Certain frozen vegetables in boilable pouches take *longer* to cook than if you had prepared the fresh counterpart."

While plastics currently used in cooking bags and pouches may be safer than the early ones, unknown and unanticipated interactions may result when various substances come in contact with food. The more complex the materials, the greater the possibility for interactions. Plastics *are* complex, and have a record of leachability.

Now that the FDA considers plastic bags and pouches regulable as food additives, a manufacturer must petition the agency for permission to use the article with the type of food to be cooked, the cooking method, and so on. Plastic bags must be confined to the specific purpose for which they have been designed.

If you like the cooking-bag technique but are worried about plastic, there is another option. A parchment paper product made solely of wood pulp allows food to be cooked in a bag, without contact with water or pot. Water cannot seep through the parchment, and there is no sticking or burning. If you need to avoid certain metal contacts, and if you need to, or wish to, avoid plastic contacts, parchment cooking is useful. It is also useful when camping or traveling, for you can use water from lakes or streams of unknown purity without endangering your food.

Moisten the square parchment sheet and then place the food on top; pull up the four corners and tie them securely. The food is ready to cook by lowering the parchment into a pot of water. Although the parchment will become stained from food contacts, it can be reused. The parchment sheets are called Vita Wrap®. If you cannot find them at a local store, write to the manufacturer and ask about the nearest distributor: Alpha Products, Inc., North Miami, Fla. 33181. Also, see the address for patapar paper on page 44.

PORCELAIN-ON-STEEL COOKWARE: *See* Porcelain-enamel Coated Cookware

PORCELAIN-ENAMEL COATED COOKWARE

If you like iron cookware you'll love porcelain-enamel coated cookware. This popular type of utensil, also called "porcelain on steel," "vitreous enamel," "porcelainized," and "ceramic coated," combines a good heat-conducting metal with an easily cleaned enamel surface. The porcelain has a nonstick quality. As long as the coating remains in good condition, the surface is durable and no metal leaches into food. The rock-hard finish will not scratch, rust, fade, or peel. The only drawback is that, like cast ironware, these utensils are heavy.

Porcelain-enamel coated cookware consists of a specially formulated highly durable glass fused to preformed metal, usually cast iron, steel, or aluminum. The molten glass and red-hot metal are bonded together permanently.

The technique of making porcelain enamel is very old, with artifacts displayed in museums dating as early as 500 B.C. The technique was probably developed in ancient Egypt and the Middle East, the glass originally being applied to metal art objects such as jewelry. Much of Cleopatra's jewelry was probably porcelain enamel, with the metal components of gold, silver, copper, or bronze. Think of Cleopatra as you swing a porcelain-enamel coated utensil from shelf to stove!

The durability of porcelain enamel accounts for the survival of so many artifacts. Several years ago, a brilliantly colored Celtic shield of porcelain enamel on bronze was dredged from the mud in the Thames River. The shield was dated from the Gaelic Wars, yet its colors and finish were still clear.

Although the technique of making porcelain enamel is very old, its application to cooking utensils began in 1830, when an Austrian craftsman, Bartelmes, produced the first porcelain-enamel coated pot by dusting glass particles onto red-hot iron. By 1850, enameled kitchenware made with sheet iron was produced in Austria and Germany, and, by the late 1800s, became popular in America.

Some cookware, resembling porcelain-enamel coated products in appearance, are made of synthetic enamel, and called "baked enamel." These coatings, made of resin-based paints, are applied to metal and are either air dried or baked on with mild heat (200° to 400° F.). Such coatings, of plastic acrylic and polyamide, are not known to be dangerous, but are softer than porcelain enamel. They scratch and stain readily. Since the appearance of resin-coated and porcelain-enamel cookware may be similar, read labels carefully.

Quality of enamel varies with price. If you invest in this cookware, buy the best. Consider choosing colored enamel, which can be an energy saver, since a dark surface will absorb more heat with top-of-the-stove cooking, and you can generally lower oven temperatures for baked goods by 25° F; for meats, from 25° to 50° F.

Porcelain enamel, which resists heat and alkali detergents, is dishwasher safe. However some *covers* sold for these utensils are painted, and may *not* endure harsh dishwasher treatment.

Use warm soapy water to wash porcelain-enamel utensils by hand. A preliminary soak for broiler pans minimizes scrubbing. To clean baked-on grease, use a steel-wool pad, or occasionally, cleansers. Remember that abrasive cleansers used repeatedly or too vigorously may eventually scratch the glaze. Although the coating is still protective, the utensil then becomes much harder to clean.

PORCELAINIZED COOKWARE: *See* **Porcelain-enamel Coated Cookware**

RESIN-COATED NONSTICK COOKWARE: *See* **Nonstick Cookware**

SILICONE-COATED NONSTICK COOKWARE: *See* **Nonstick Cookware**

STAINLESS STEEL COOKWARE

You may enjoy the lasting attractive appearance of stainless steel cookware. Introduced in 1938, these utensils have an attractive finish that will not corrode or tarnish permanently, and their hard, tough, nonporous surface resists wear. Once stainless steel has been stamped, spun, or formed into utensil shape, it takes an extremely hard blow to dent it.

Like other steels, stainless steel is an alloy, made from a combination of iron and other metals. What makes it different from other steels is that it contains at least 11 percent chromium. It is the chromium that makes the steel relatively stainless all the way through. Stainless steel may also contain other metals, such as nickel, molybdenum, columbium, or titanium. These metals can contribute special qualities such as hardness, resistance to high temperature, and resistance to scratches and corrosion.

The main disadvantages of stainless steel cookware are that steel is a poorer conductor of heat than other metals and that it fails to heat evenly. To minimize these drawbacks, most stainless steel top-range cookware of good quality is combined with other metals, usually aluminum, copper, or carbon steel to improve heat conductivity.

Two-ply utensils commonly have a stainless steel interior with another exterior metal. In some instances, this arrangement is reversed, with the stainless steel outside.

Three-ply utensils have stainless steel on both the inside and outside surfaces, with a layer of copper, carbon steel, or aluminum forming the core.

Bottom-clad utensils are formed with solid stainless or three-ply, and copper is plated to the bottom or aluminum is applied to the bottom by casting, bonding, or metal spraying.

Five-ply bottom-clad utensils are made by the three-ply process, adding two layers on the bottom. These utensils are made with stainless steel on both the inside and outside surfaces, and three layers of aluminum forming the inside core. Five-ply bottom-clad stainless steel utensils are heavy, durable, and the most costly of various types of stainless steel pots and pans.

Stainless steel bakeware, intended for oven use, is usually made of solid stainless steel.

You may have the impression that stainless steel is inert and will not leach metal into food. While stainless steel is *relatively* inert compared to other metals, it is not entirely inert. Also other metals present in the alloy can be released from stainless steel cookware. Although the levels may be exceedingly low, and not considered hazardous to the average person, they may induce adverse effects in individuals who may be extremely sensitive to specific metals. One patient's symptoms of sensitivity to nickel were attributed to the extremely low level of nickel in a stainless steel pin that had been used surgically to repair a bone in his body. Another patient proved to be sensitive to an extremely low level of copper present in an IUD. Pots and pans should also be considered as factors when individuals are found to be sensitive to metals.

You may have the idea that stainless steel is "stainless." Actually, stainless steel cookware may discolor in spots, but such discoloration does not affect the utensil's performance. High heat may cause a mottled, rainbowlike discoloration commonly called "heat tint." Cooking certain starchy foods such as rice, potatoes, or peas may cause a stain on the inside of the pan. Both heat tint and starch stains can be removed with stainless steel cleaners. Undissolved salt will "pit" steel surfaces. Do not allow salty or acidic foods to remain in stainless steel utensils for long periods of time.

Buy good-quality, heavy-gauge stainless steel cookware made by reputable manufacturers. Check that covers fit tightly and that handles are functional and safe. Choose pieces that are clad with another metal on the bottom to assure fast, even heat. Cheap utensils of thin-gauge stainless steel do not have metal-clad bottoms and food burns very readily in them. You will waste a great deal of time and energy scouring burnt utensils.

Buy only the pots and pans that you use in your style of cooking, and in sizes that fit your needs. You can always add more as you need them. Give thought before buying an entire set. Resist high pressures from door-to-door salespeople or mail-order companies (both deal almost exclusively with sets made of stainless steel) to purchase entire sets. Sets of cookware are exceedingly costly, bulky to store, and probably contain some pieces that you will not use.

Keep stainless steel cookware clean with soap and water. If any scouring is necessary, use a soft cloth and the least abrasive cleanser. After long use, and especially after repeated harsh scourings, nicking and pitting of stainless steel may expose metals from the inner layer. If this happens, discontinue using the utensil with food.

STONEWARE: See Clay Earthenware

TEFLON COOKWARE: See Nonstick Cookware

TINWARE

The use of tinware can be traced back to ancient times, when tinplate was used commonly in Egyptian kitchens. Phoenician sailors ventured to the Isles of Tin, now part of Great Britain, to obtain tinplate for Egyptian use. Today, tinware is still important in home kitchens as well as in commercial bakeries. Many of our muffin forms and baking pans are made of tin-plated steel.

Tin plating by plunging plates of iron into molten tin was perfected in Germany in the sixteenth century. By about 1670 the use of tin as a protective coating for metal was brought to England.

Modern tinware, consisting of steel plated with tin, is durable, relatively inexpensive, and has excellent baking qualities. Very little care is required and it is highly resistant to denting and scratching. Tinplate provides the necessary protection that helps the steel resist rusting and staining, and is entirely safe when used in this way.

VITREOUS ENAMEL COOKWARE: See Porcelain-enamel Coated Cookware

2

Kitchenware

Some of the most troublesome [environmental] exposures have not been adequately described and there is no general knowledge of their potential hazards. The chief reason for this is that these materials have become integral parts of our current existence. Not being readily avoided accidentally, they are not usually suspected. Not being suspected, they are not usually avoided deliberately.
—THERON G. RANDOLPH, M.D., 1962

Look around your kitchen. You probably have utensils, cutlery, and cutting boards of various substances, including metal, decorated glass, plastic, and wood. What are the shortcomings as well as the desirable features of such substances?

An alphabetical listing of various materials used in kitchenware follows, with an indication of the advantages and shortcomings of each.

DECORATED GLASSWARE

If you have young children at home, you probably have some brightly colored decal-decorated glasses. Although the decals appeal to young children, how safe are they on drinking glasses?

This question was raised in 1977, after the McDonald's fast-food chain, in a promotional drive in New England, made available drinking glasses with decal cartoons. The Massachusetts Health Commissioner was alerted when high levels of lead were found in the decals. The commissioner urged parents to prevent children from using such glasses, and to place the glasses out of children's reach. The commissioner also urged a ban on further distribution of the glasses. Similar actions followed in other New England states.

The FDA investigated and confirmed the potential hazard. The agency advised that the glasses should not be used, and admitted that lead from the decal could migrate, especially in contact with acidic fruit juices, or when exposed to and washed with chemicals for a long time. But the FDA did not recall any glasses. The agency found "no evidence [that] the lead can contaminate the liquid inside," and judged that no "acute health hazard" existed.

Massachusetts officials disagreed, and contended that the lead was very close to the glass rim and could be dangerous to children who might chew it off, or handle glasses with sticky fingers and then lick their fingers.

In response, the FDA and EPA urged parents to prevent children from licking or gnawing at the outside of the glass. This recommendation was made *after* the FDA had told parents that the glasses should not be used. FDA's inconsistency reflected the agency's lack of certainty as to what measures needed to be taken.

The problem was far greater than the few million glasses distributed by McDonald's. The glass industry, using the same process of heat-fusing ceramic designs, annually produces some 400 million glasses. (Glasses decorated with gold or other precious metals are not manufactured by this method.) During the McDonald's flap, the glass industry defended the safety of such products. The process is worldwide in scope and has been used for the last fifty years. Billions of decorated glasses have been manufactured, and doubtless many are still in use.

High levels of lead can cause irreparable physical and mental damage, and even death, in children under six years. As a result of the McDonald's controversy, the Childhood Lead-Poison Prevention Program of Massachusetts asked the EPA's regional laboratory in Massachusetts to run tests on the McDonald's and other decorated glasses.

Dr. Thomas M. Spittler, who ran the tests, found that the new, unused glasses distributed by McDonald's, contained no detectable lead in the glass. But in the decal, depending on the color, the lead content went from 7.4 to 10.1 percent by weight. On older glasses that had been used for three to five years, the decorations had become chalky and could be easily removed, say, by scraping them with a fingernail. By rubbing the decal for only five seconds with a cotton swab dipped in fruit or vegetable juice, as much as twenty-six micrograms of lead could be removed. More tests with older glasses showed that about 15 percent of the lead had already been removed. The exposed lead on the worn surface could be leached by acids commonly found in fruit juices.

"Young children, my own included, will frequently lick or even scrape their teeth on painted surfaces when drinking," Spittler said. To learn if this habit was apt to remove lead in measurable quantities, Spittler placed three drops of V-8 juice on his tongue, licked the surface of an older glass for five seconds, rinsed his mouth with distilled water, and analyzed the washing. He repeated the test. Each time, about five micrograms of lead was removed from the glass. Other tests showed that glasses, subjected to repeated dishwashing, released increased amounts of lead from the decal.

No federal standard for lead in glassware existed. Where state tests exist, such as in Massachusetts, they are grouped with other ceramic materials. When conducted, leaching tests are only performed on the *inside* of the container where the food or beverage is expected to be in contact. Yet Spittler's findings raised an awareness that the *outside* should also be examined for health hazards.

Spittler's evidence demonstrated the need for further testing of decorated glassware. Recognizing the inadequacy of public protection, the FDA, CPSC (Consumer Product Safety Commission), and EPA organized an interagency task force to evaluate the safety of glassware decorations. Cadmium as well as lead would be considered, since cadmium, too, can leach from glassware, although only at about 1/25 the rate of lead.

Early in 1978, the task force reported "substantial increase above normal daily intake of lead and cadmium might result from the routine use of certain types of glasses and indicates that some glassware represents a hazard." Tests with nine types of decorated drinking glasses, and simulating child use with licking and gnawing, had released dangerously high levels of lead and cadmium from two types; one type was doubtful and the remaining six leached low lead levels not considered hazardous.

In December 1978, the task force recommended a voluntary standard be established to control lead and cadmium on decorated glassware, and the standard became effective in March 1979. If industry failed to comply voluntarily, further government action would be taken. The group recommended that lead leached from the lip and rim of decorated glasses not exceed 50 ppm, and leached cadmium, 3.5 ppm. The task force concluded that a standard was necessary for the lip and rim area of the decorated glassware, with the risk confined to the top 20 millimeters. In addition, good quality control of techniques used for decorating glassware could reduce the risk of leached toxic metals from all portions of the glass.

Although the McDonald's glassware incident received prominent press coverage, other types of hazardous drinking containers seized by the FDA show that the problem continues. In one instance, a Long Island resident complained to the FDA's New York District that some of the luster from his silvery, mirror-finish bar glasses disappeared after his drink remained in a glass for half an hour. Laboratory tests showed that the solution applied to the glassware contained an unacceptably high level of nickel which dissolved from the glass into the drink. The FDA learned that the manufacturer had received other consumer complaints about the product. The agency withdrew nearly 10,000 of these glasses from the market.

It is probably prudent to discontinue using any glassware with decals. Even if the decoration is low on the glass, away from the rim, there is no assurance that lead and cadmium, which gradually leaches out, will not contaminate *other* food utensils in your dishpan or dishwasher, or that a child will not lick or scrape it off. If you enjoy decorated glassware, consider glassware decorated with gold, silver, or other precious metals. Acid-etched glass and cut glass are other options.

DINNERWARE

Since you use dishes several times each day, it is especially important that your set is sound and safe. Cracked or crazed dinnerware can retain bits of food and become contaminated with microorganisms. Poor glazes can leach toxic metals. And what about other types of dinnerware, including wood, metal, and plastic?

Generally, chinaware and porcelainware are sound and safe. But from time to time, the FDA has made seizures, usually of inexpensive imports with poor glazes. In 1977, for example, the agency seized nearly 6,000 dozen porcelain cups and saucers from Hong Kong which contained high levels of leachable lead.

Earthen dinnerware is generally sound and safe, but may offer the same hazards as chinaware and porcelainware when the glazes are poor. In 1977, for example, the FDA seized one lot of 180 dozen earthenware mugs imported from Japan which contained high levels of leachable lead.

Wooden plates may be attractive, but they are impractical to use with hot food. The finish becomes worn, and plates will crack and retain bits of food.

Pewter dinnerware is attractive, too. Pewter is an alloy, its major metal being tin. Since the composition of old pewter is uncertain, and may contain lead, it should not be used in food contacts. New pewter is lead-free, but may leach other metals such as antimony at low levels.

Plastic dinnerware can be made with melamine-formaldehyde resin. The FDA allows the resin in food contact, provided that no more than three moles of formaldehyde react with one mole resin in a water solution. (A mole is the quantity of a chemical substance that has a mass unit weight equal to molecular weight.) The FDA acknowledges that formaldehyde leaches into food but considers the level "insignificant." Resins used in plastic dishes may be mixed with refined wood pulp, lubricants, and polymerization-reaction control agents. Interactions between components in plastic dinnerware and foods can be seen readily in strong, hard-to-remove stains, such as coffee or tea in plastic cups.

Also, your plastic cup may disintegrate, if you take lemon with hot tea. A consumer experienced such a vanishing plastic cup, and complained to the state of Connecticut. Tests confirmed that the plastic could be attacked by its contents, especially lemon. Investigators experimented by placing the cups in lemon oil and warming them. Some pieces of the cups dissolved completely.

Use sound dinnerware with good glazes. Discontinue using any that show cracks or crazes in the glaze. If you plan to purchase new dinnerware, invest in good-quality products, manufactured by reputable companies. Inexpensive imports of unknown origin may carry some hazard. Restrict use of wooden and pewter plates to decorative use *under* bowls or smaller plates. Be skeptical about plastic dinnerware, especially if it stains readily or gives off odor.

METALWARE

Look around your kitchen and check any metalware. Some may be exposing your food to toxic metals, others to metals considered safe in small amounts but toxic at high levels. Lead and cadmium are examples of the former, and zinc an example of the latter.

Lead is a toxic metal to be avoided whenever possible. Do you have silver-plated hollowware goblets, baby cups, coffee sets? If the plating is worn or cracked, hollowware dissolves some lead under the plating, especially if it is in contact with acidic food or liquid.

Cadmium is another toxic metal you should avoid as much as possible, and, like lead, it is especially hazardous in contact with acidic food or liquid. During World War II, when aluminum was scarce and other metals were used as substitutes, cadmium was used to manufacture some serving trays. The entire crew of a battleship was inactivated, not by the enemy, but a quantity of fruit gelatin prepared and served in cadmium trays.

The accessory ice trays in home refrigerators may be cadmium coated. Although such trays are probably safe for making ice cubes, never use them for freezing sherbet, which is acidic and will leach cadmium from the trays.

Your kitchen offers numerous exposures to cadmium. Some are avoidable, and some beyond your control. Whatever you can do to minimize cadmium intake is desirable, since cadmium accumulates in the body. You can avoid exposure to cadmium . . .

from cadium-plated roasting pans—cadmium is dissolved by the fat in roasted meat or poultry;
from cadmium in silver polish products that clings to poorly rinsed polished kitchen utensils;
from cadmium residues in some foods, absorbed from some plasticizers used in plastic wrappings.

You'll find it harder to avoid other cadmium exposures unless you are willing to avoid certain foods, beverages, or containers. Cadmium residue is found in:

- butter, leached from galvanized milk cans into cream
- olive oil, from presses and metal containers in which the oil is processed and stored
- processed meat, from contact with processing machinery
- cola beverages, instant teas and coffees, both regular and decaffeinated, from processing
- dried and canned fish, from smoking and canning processes, and possibly from contact with galvanized wire netting
- soft water, from galvanized iron or copper pipes
- sugar, from the refining process
- tin and aluminum cans, from recycled metal

Zinc, safe in small amounts, becomes toxic at high levels and is made even more hazardous by its contamination with toxic metals including cadmium. Cadmium levels in zinc may be as much as 1.5 percent. Other contaminants include antimony, arsenic, and iron.

Galvanizing with zinc is an inexpensive way to give a protective coating to other metals, especially iron. Before the era of plastics, for example, water buckets for scrubbing were usually made of galvanized metal. Look around your kitchen and check on galvanized containers. Galvanized metal containers, such as milk cans, are popular for bulk storage. Improper use of galvanized containers for food preparation and storage, uses for which they are not intended, has caused many cases of metal poisoning, and is a constant source of worry to health authorities. Some examples:

In 1937, soldiers suffered severe gastrointestinal distress and diarrhea after drinking limeade prepared in new galvanized iron cans intended for garbage disposal. Was the poisoning from high levels of zinc or other metals? A limeade prepared in a similar manner and analyzed was found to contain nearly fifty times the average daily zinc intake. Dissolved antimony was also found in the limeade, and it had converted to a tartrated form which nearly tripled the dose and raised it to an emetic level. Both zinc and antimony were judged responsible for the limeade's toxic effects.

In 1964, two incidents of mass food poisoning in California were attributed to zinc. In one instance, galvanized tubs were rinsed with acid, then stored with chicken and spinach. The rinsing released a high concentration of soluble zinc salts into the stored food. More than three hundred people became ill while attending a celebration. In the second incident, an alcoholic punch, stored more than two days in a galvanized container, made forty-one persons ill. Analysis showed that a five-ounce portion of the punch contained enough zinc to be within the emetic range.

Check the plating on metalware and make certain it is sound. If it is worn or cracked, either have it replated or discontinue using it with food. If you don't know the composition of the accessory ice trays or roasting pan, but suspect that they are cadmium coated, contact the manufacturer. Or, replace them with trays and pans made of other metals. If you use silver polish, make certain to rinse the utensils thoroughly. Since it is impossible to know which types of plastic wrappings may contain cadmium residues, choose the safer types of nonplastic food wrappings. Don't use galvanized metal containers to prepare or store foods or liquids.

STORAGE CONTAINERS

What kind of containers do you prefer for food storage? If you are a typical American, your first choice is glass. Although many glass containers have been replaced by plastics, when consumers have been asked about their preference, they have overwhelmingly favored glass. Glass keeps food fresh, retards spoilage, and imparts no flavor to the food.

Glass storage containers have been replaced by plastics so largely, that I purchase suitable old glass containers in junk shops whenever I find them. Formerly, they were both common and inexpensive, selling for ten or fifteen cents in five-and-ten-cent stores. Now, as "collectibles," they bring as much as six to eight dollars in some places as "Depression glassware"!

Plastic storage containers are commonplace, and popular especially for storing home-frozen food. The quality of plastic storage containers has been improved through the years. Early plastic was unsatisfactory, as formaldehyde was used in processing and leached from the plastic. In time, the plastic deteriorated, with off-odors, discoloration, and tackiness. With improved techniques, these problems have been eliminated. Usually storage containers are made of durable polyethylene. Most people find them convenient and satisfactory. However, individuals who are extremely sensitive to plastics of any type may need to avoid them and use glass exclusively.

If you prefer glass, save empty containers from peanut butter, honey, fruit juice, and so forth, since it is becoming more difficult to find glass storage containers. Before storing food in either glass or plastic containers, allow it to cool; hot food may crack glass and melt plastic. Allow an inch headspace before freezing food or liquid in glass or plastic containers.

STORAGE WRAPPERS

Food wraps, once made basically of paper, now include a wide variety of materials and characteristics. They may be waxed, greaseproof, moistureproof, plastic coated, or silicone treated, and made of materials such as glassine, kraft, and vegetable parchment. Although cling-wrap plastics have replaced many paper-wrap products, household paper wraps are still readily available in food stores.

The wax used in some paper wraps may be treated with an antioxidant, a volatile substance added to prolong the storage life of the food. While the presence of antioxidants is printed on the labels of foods such as packaged dry cereals, it is not printed on containers holding rolls of wax paper intended as food wrap. It may be necessary to contact the manufacturer of the product to learn whether the wax in the paper has been treated. This information may be important for individuals who are sensitive to the antioxidants.

Shun pesticide-treated shelf-liner paper for use in your kitchen cabinets, on pantry shelves, in dresser drawers, or elsewhere in your home. Pesticides are toxic and volatile.

Freezer paper is bleached, white kraft paper laminated to cellophane. It appears to be well tolerated by individuals who react adversely to some plastics.

Plain cellophane, the first transparent flexible food wrap, has been in use since 1924 when it was first introduced by Du Pont. Although cellophane's main ingredient is cellulose, modern cellophane may be plastic coated to eliminate brittleness at low humidity and temperature. Even those individuals who appear to be sensitive to certain plastics seem to be able to tolerate modern cellophane despite its plastic components. You can transfer food wrapped in packaging materials of unknown identity into cellophane bags for storage at room temperature or in your refrigerator or freezer.

Hypoallergenic cellophane bags are available from Lange's Nu Vita Foods, Inc., 7524 SW Macadam Avenue, Portland, Oreg. 97219, or from S. Freedman & Sons, Inc., 7845 Old Georgetown Road, Bethesda, Md. 20014, for cellophane bags or cellophane film wrap.

You can obtain patapar paper, for cooking or storing food, from Patapar World Organic, P. O. Box 8207, Fountain Valley, Calif. 92708.

Metal foil is a convenient and protective food wrap. While lead, tin, and zinc foils are mainly of historic interest, aluminum foil is common and versatile. It resists grease, is odorless and tasteless, does not shrink, swell, soften, or discolor. Aluminum foil is a good heat conductor but does not burn nor crack when used to wrap and store foods in the freezer. It is a better barrier against air and moisture than many plastic wraps or waxed papers.

Since aluminum foil can be pressed into many shapes, it can serve as lids for sealing containers tightly, yet peel off readily. In the oven, aluminum foil can line a broiling pan or catch drippings under the pan. Or it can be folded and shaped into a vessel to be used for oven cooking.

"Heavy duty" or "heavy weight" aluminum foils are completely impervious to air and moisture.

You can reuse aluminum foil. However, after it is creased it may lose some of its imperviousness, and it also tends to develop pinholes.

Be aware of some negative qualities of aluminum foil. Don't use it to wrap cured or salty meats, for salt will corrode the foil. Don't use it in direct contact with acidic foods such as citrus fruits, berries, tomato sauce, pickles, relishes, sauerkraut, jellies, jams, fruitcakes, and fruit pies, for the acid in these foods may create pinholes in the foil. Don't attempt to cover the entire oven rack or line the entire oven bottom with aluminum foil. The heat concentration could harm the oven. Also, the food will cook unevenly.

If you are among those who believe it is dangerous to use aluminum foil for cooking, be relieved to learn that this wrapping offers benefits. For example, wrapping meat in aluminum foil before grilling can prevent deposits of carcinogenic polycyclic aromatic hydrocarbons from charcoal on the meat's surface. (See Charcoal Grills.) Also, aluminum foil can conserve nutrients. Potatoes, wrapped in aluminum foil before baking, retain more ascorbic acid than potatoes baked without wrapping.

If you enjoy the convenience of aluminum foil but still have qualms about using it in contact with food, you can package food in wraps of your choice and use the aluminum foil as the outer wrap.

Plastic food wraps have also become very popular. The raw materials of each plastic differ, and some wraps may contain several plastic combinations. Manufacturers and the FDA agree that there is no practical way of simply looking at a plastic wrap and determining its composition. The certain means is by costly analysis in a laboratory.

From all the testing done to date, polyethylene, one of the commonest of plastic food wraps, appears to be safe. Make certain, however, that you are buying a polyethylene product without polyvinyl chloride (PVC), a plastic of questionable safety. In the article "Around the House" (*Good Housekeeping*, March 1976), readers were told that polyethylene plastic food wraps do *not* contain PVC. Infraspectroscopy tests conducted at Worcester (Mass.) Polytechnic Institute, however, demonstrated that some of the polyethylene plastic food wraps did contain PVC. Samples that were *free* of PVC were: Handiwrap (Dow); Staff Plastic Wrap (Staff Supermarkets); First National Clear Plastic Wrap (First National); Snap-off Utility Bags (Union Carbide); Shake 'n Bake (General Foods); Baggies (Colgate-Palmolive); Finast Bread Wrapper (Finast);

Stop 'n Shop Bread Wrapper (Stop & Shop); and Nancy Martin Bread Wrapper.

Since manufacturers may change formulations, presently available products do not necessarily have the same composition as those cited above, tested in 1976. For specific information, look on the package. The label of Gladwrap, for example, specifically states that the product is made of polyethylene; that is, it does not contain PVC. If this information is not supplied on the package, contact the manufacturer. At the same time, suggest that this information be added to the package to make it available to all consumers.

WOODENWARE

What woodenware do you have in your kitchen? Cutting and pastry boards, kitchen countertops and chopping blocks, chopping bowls and salad bowls, salad serving implements, butter and cheese knives, dip spreaders, egg cups and egg spoons, rolling pins, and numerous types of stirring spoons have traditionally been made of wood. You may enjoy the tactile pleasure of smooth-textured wood as well as its attractive graining. Wood is not cold to the touch, and it does not contain leachable metals. Yet woodenware has certain drawbacks.

In time, the coating used to seal wood surfaces may wear off, peel, or chip, and contaminate food. Uncoated wood is potentially hazardous, for food may be retained in the porous wood and become contaminated by microorganisms. Also, wood exposed to temperature changes can crack, and food becomes lodged in the crevices.

You need to handle wooden cutting boards with particular care. Under the present federal/state meat-inspection regulations, you have *no* assurance that inspected raw meat or poultry is free of pathogenic microorganisms. For example, trichinosis organisms may be present on inspected raw pork, or salmonella organisms on inspected red meats and poultry. The same hazardous situation is present with raw fish, inspected by other agencies. The official view is that such foods will *become* safe when you cook them; what officials ignore is the hazard of handling these foods while they are still raw. If you prepare raw meat, poultry, or fish on a wooden cutting boards. To eliminate the possibility of cross-contamination, carrying organisms which may be present in these foods. If you then slice bread or prepare fruits and vegetables intended for salad on this same board, these foods may become contaminated if you have washed the board inadequately or if the board has cracks.

Check sealed surfaces of woodenware and replace those where the finish shows signs of wearing off, peeling, chipping, or cracking. *Scour* the surface of a wooden cutting board after you handle any raw flesh foods, and scrub your hands and cutting implements, too. Discard cracked cutting boards. To eliminate the possibility of cross-contamination, reserve one cutting board exclusively for the preparation of raw flesh foods, and use another for other foods. An extra cutting board is a worthwhile investment. Or use an impermeable glass board, such as the Corning portable counter saver. Scour it carefully after each use.

PART II

SOME COOKING DEVICES

3

Microwave Ovens

Although ionizing radiation seems to loom largest as a hazard, it would not surprise me in the least if non-ionizing radiation were ultimately to prove a bigger and more vexing problem.

—PROFESSOR CHARLES SUSSKIND
Department of Electrical Engineering
University of California, 1968

Are you among the 10 percent of all Americans who, by 1977, had installed a microwave oven in your kitchen? If not, will you be among the 25 percent who, if predictions prove correct, will have purchased one by 1980? Whether you already own and operate a microwave oven or plan to buy one, you should know its undesirable features as well as its benefits.

A microwave oven cooks food by penetrating it with radiation. This immediately generates heat within the food by agitating its molecules. The air surrounding the food is not heated. In traditional cooking, the air surrounding the food needs to be heated before the heat can penetrate the food. With microwaves, it is possible to heat the food far more rapidly than in traditional cooking.

A microwave oven is convenient for thawing and cooking frozen food, and since the oven remains cool, food does not spatter onto its interior while being cooked. This makes the oven very easy to clean.

A microwave oven also helps to retain nutrients in food. In vegetables, ascorbic acid is retained as well in microwave cooking as when they are carefully cooked in minimal amounts of water. In some cases, microwave cooking helps retain vitamins even *better* than traditional cooking methods.

Despite the speed, convenience, and nutrient retention, there are drawbacks. The microwave oven is not an all-purpose cooking appliance. It can supplement a conventional range which has both an oven and top-stove elements, but does not replace it in versatility. Unless the microwave oven is equipped with a special unit, for example, many foods will not brown when cooked.

Although microwave ovens have been touted as energy savers, their economy has been grossly exaggerated. The National Bureau of Standards reported that microwave ovens are about 40 percent efficient. This percentage is good compared to 14 percent efficiency for electric ovens and only 7 percent for gas ovens. However, it was found that the microwave oven saved only about $10 a year on fuel bills for the average household, and even less if the oven was used to thaw frozen foods. Considering that a microwave oven costs about $200 more than a conventional oven, it would take quite a while to write off the additional expense.

A misconception persists that food cooked in a microwave oven becomes radioactive. Not so. Microwaves are a non-ionizing form of radiation; that is, they lack the ability to create ions such as those associated with radiation from radium, X-rays, or nuclear weapons.

Nevertheless, the non-ionizing radiation needs to be studied, for it may adversely affect human health. Our present understanding of the biological effects of non-ionizing radiation is quite incomplete, but research suggests that microwave radiation may have fundamental and profound effects on humans at levels that are as yet undetected. Various studies in the Soviet Union have demonstrated that chronic exposure to low levels of microwave emissions may result in many biological changes, including a decreased rate of heart muscle contraction; low blood pressure; changes in the blood's composition; increased activity and enlargement of the thyroid gland; hormone imbalance; alteration in the central nervous system, brain wave patterns, and behavior; and a wide range of nonspecific ailments. In view of these findings, the U.S.S.R. established permissible exposure levels that are *a thousand times stricter* than those proposed in the U.S.

In general, American scientists have viewed the U.S.S.R. reports skeptically and have conducted few experiments. The few American scientists who have investigated the effects of chronic human exposure to low levels of microwave radiation tend to give credence to the Soviet findings.

Microwave ovens have had a long history of radiation leakage. A 1969 survey of microwave ovens in home use in New York, New Jersey, Massachusetts, and Mississippi showed that one-third of the units exceeded the industry's voluntary standard. Similar results were found in a Florida survey. Studies of home microwave ovens in two other states and in the District of Columbia showed that even when the oven doors were well closed, up to 25 percent of the units leaked radiation in excess of the industry's voluntary standard. Of thirty ovens delivered to the Walter Reed Army Medical Center, twenty-four had to be rejected due to radiation leakage.

These and additional findings prompted a nationwide government-industry survey of microwave ovens in early 1970.

Also, the U.S. Surgeon General formed an Ad Hoc Task Force on Microwave Ovens to identify brands and models that leaked excessively, so that manufacturers could repair them. This survey showed that 10 percent of all ovens leaked radiation above the industry's voluntary standard.

The main problem was faulty design of the safety interlocks that keep the oven door closed while the microwave oven is in operation, as well as improper maintenance and servicing. Users were cautioned to keep at least an arm's length away from the front of the oven while it was in operation; to switch off the unit before attempting to open the oven door; and not to allow children to use the viewport to watch foods being cooked. The viewport contains a wire mesh screen which blocks microwaves but allows the cook to peer into the oven. However, if the mesh is damaged, dangerous emissions can penetrate the viewport. Newer regulations require that such wire screens be protected by heavy glass or plastic.

In May 1970, the FDA proposed safety standards for microwave ovens and limitations of the power density of microwave radiation. The standards were to be applicable to all ovens manufactured after October 6, 1971, but older microwave ovens were to be covered by the "defect provisions" of the standards. Safety requirements were specified for the door interlocks. Despite attempts to tighten safety, the leakage problem persisted due to the difficulty of adjusting safety interlocks which operated from the door's motion.

In April 1973, the FDA proposed tighter controls. It would require *two* safety locks, with one concealed so that it could not be operated manually.

While the FDA wrestled with these safety problems, Consumers Union (CU) brought to public attention the problem of hazards in microwave ovens. CU tested fifteen popular countertop microwave oven models and concluded that they were not completely safe. Although all of the models met the emission standards established by FDA's Bureau of Radiological Health (BRH), CU contended that "until much more evidence is available regarding the safety of low-level microwave radiation, we do not feel we could consider a microwave oven 'Acceptable' unless there is no radiation leakage detectable." CU designated all fifteen models as "not recommended."

Since passage of PL 90-602, microwave-oven manufacturers have tried to make their products safer. But defective microwave ovens are still a problem, and recalls are made from time to time. And although the leakage problem has been reduced, it has not been eliminated. Manufacturers are required to report any injuries to users of microwave ovens to the BRH.

In 1977 in Kentucky, two waitresses sustained burns from heating foods in a microwave oven in a fast-food restaurant. The women reported feeling tingling sensations as they removed the food from the oven. Later, they experienced pain, swelling, and discoloration in their hands. BRH examined the oven but was unable to find any defect.

Following the Kentucky incident, it was learned that in the previous year a man had filed a legal claim for damages from burns inflicted while removing food from a microwave oven in a restaurant elsewhere. According to his testimony, the oven failed to cease operating after the door was opened. The man's hand swelled so much that amputation was considered. After the incident, the oven was tested but it failed to continue operating after the door was opened. The restaurant had not reported the incident to the BRH since the injury had not been considered a legitimate claim.

If you already own and use a microwave oven, check to make certain it is in good working condition. The interlock system should function properly. If the oven was manufactured before 1971, check the wire mesh screen over the viewport and make certain it is still sound. If the oven was manufactured after 1971, the viewport is probably covered by heavy glass or plastic. Check the soundness of this cover. If you have any suspicion that the oven is malfunctioning—a faulty door seal or failure of the oven to cease operating after the door is open—stop using it. You can do one of two things. Contact the manufacturer to have it checked and, if necessary, repaired. Or contact the Bureau of Radiological Health, FDA, Washington, D.C. 20204, and inquire how the oven can be checked for defects or possible radiation leakage. Do not rely on numerous devices sold as do-it-yourself tests to check the safety of microwave ovens. The FDA has cautioned that such devices are "inaccurate and unreliable."

If you do not own a microwave oven but are thinking of buying one, do some soul searching. Judge the slight benefits of speed and convenience against the potential personal risk of radiation leakage and the larger problem of contributing to the total environmental burden of microwave emissions. What does your conscience dictate?

4

Charcoal Grills

It is safer to boil food or to poach food than to charcoal broil it. We have evidence that in broiling food we form the cancer-producing substance in the process of cooking.
—DR. ARTHUR C. UPTON
Director of the National Cancer Institute
radio broadcast, Christmas 1977

Dr. Upton's message may have dampened holiday cheer for many Americans. The backyard charcoal grill was considered as American as apple pie. The basis for Upton's warning was well known to cancer researchers, but not as yet to the American public.

Upton noted that at least two kinds of substances are formed in charcoal broiling: polycyclic aromatic hydrocarbons (PAH), known to be carcinogenic, including benzo(a)pyrene (BP), related to the tar in the condensate from cigarette smoke; and other carcinogenic substances produced from the breakdown of some amino acids in the protein of the meat. Thus, not only does charcoal broiling coat the food with carcinogenic substances, but additional ones are produced within the food.

Upton's announcement was based on a long history of accumulated evidence. As early as 1775, Sir Percival Pott reported that chimney sweeps, exposed to carbonaceous soot, had a high incidence of scrotum cancer. A hundred years later it was shown that German workers exposed to coal tar developed skin cancer. Then when cancer was induced in animals exposed to coal tar, BP, one of its active carcinogens, was identified. Additional experiments showed that a number of PAH compounds were potent carcinogens.

These findings have implications for methods of cooking food. If the fat does not contact the heat source and the smoke does not reach the meat, the food is free of PAH. One study showed that PAH is not found in significant quantities if the meat is wrapped in aluminum foil and then cooked on a charcoal grill. Despite these findings, little was done to minimize the amount of PAH in grilled foods, and the public remained uninformed. Meanwhile, the sale of charcoal grills for home use continued, and the charcoal broiling of meat became popular in restaurants.

The coating of meat with carcinogens occurs even if the hot surface is not charcoal but is heated ceramic. However, no coating occurs to any significant extent if the heat source is *above* the meat.

How many Americans have heeded Upton's warnings? How many have discarded the backyard charcoal grill or have no longer ordered charcoal-broiled meats in restaurants?

Both charcoal and electric grills continue for sale, to consumers and restaurateurs, without health warnings. If you wish to grill meat, use lean meat and trim off any visible fat. Grill with the heat source *above* the meat. Avoid contact between the meat and any flame. If the heat source is below the meat, regardless of the type of heat, wrap the meat in foil. If necessary, lower the temperature to grill the meat, even if this requires longer cooking time. When meat is cooked above charcoal *in a vented pan*, which can catch melted fat and prevents contact with the flame, PAH doesn't form on the meat surface. All these practices will help you minimize BP coating of meat. Avoid charcoal-broiled foods in restaurants.

Publicity about charcoal-broiling hazards merely made ripples. But the following year, Dr. Barry Commoner, at the Center for the Biology of Natural Substances, Washington University, revealed that hamburgers, as prepared in many fast-food chains, produce mutagens—substances distinctly different from BP, but which also pose human cancer risks. Commoner found these mutagens in some commercial food preparations as well, including beef extract. Among Commoner's findings were that hamburgers, prepared with high heat (above 300° F.) and long cooked, in contact with metal, developed mutagenic substances, but hamburgers cooked *under* the heating elements of an electric broiler or in a microwave oven did *not* possess mutagenic substances.

Broil hamburger meat *under* a heat source, whether electric or gas, but do not charcoal broil. Or cook in a microwave oven, but not on its "browning plate."

5

Food-smoking Chambers

One obvious means of introducing polycyclic aromatic hydrocarbons (PAH) into food is by smoking, which until recently was carried out by exposing the food to wood smoke in a smoke oven. The few investigations that have been made of smoked fish . . . have revealed the presence of small quantities of the carcinogenic hydrocarbon benzo(a)pyrene. The relatively higher incidence of gastric cancer in Northern Russia and Iceland has been related to the large quantity of smoked fish eaten by inhabitants of these regions.

—WILLIAM LIJINSKY, PH.D.,
and PHILIPPE SHUBIK, M.D., 1965

Do you, like many people, enjoy eating smoked meat, fish, fowl, and cheese? And have you invested in a food-smoking chamber so that you can smoke foods at home? Or are you thinking of buying such a device? If so, acquaint yourself with the facts about smoked foods.

Like charcoal broiling, the smoking of food introduces carcinogenic compounds onto the treated foods. PAH compounds are deposited and then absorbed into the products during the process of smoking and storing. More than twenty-five PAH compounds have been identified in wood smoke. In addition, there are acids, carbonyls, phenols, and some forty additional substances known to be present in the smoke but not yet identified.

Even low levels of PAH compounds present in smoked flesh foods are considered to be potential human health hazards. While the level of PAH in smoked food is far lower than in charcoal-broiled food, *any* level is considered to be hazardous.

The hazards of consuming large quantities of smoked foods have been investigated in various places, especially where consumption of smoked foods was suspected of being related to a high incidence of cancer.

Studies have shown that the longer smoked fish is stored, the greater the amount of benzo(a)pyrene found in the food. In freshly smoked fish, BP was stored on the scales and skins. Immediately after the fish was smoked, BP penetration was so low that it was not detectable. However, after the fish was stored, BP became detectable.

Avoid *all* smoked food, whether it is lightly or heavily smoked. By doing this, you are helping to reduce your unnecessary exposure to known carcinogenic substances.

Read labels carefully. It is possible, for example, to choose sardines that are not smoked among many brands that are "smoked" or "lightly smoked."

Don't be tempted to purchase a smoking chamber for home use, and if you already have one, discontinue using it.

Remember that smoked meat, fish, fowl, and cheese, as well as heavily roasted coffee, beurre noire, burnt toast, and vegetable oil heated to the smoke point have all developed PAH compounds and should be avoided.

There are still many good foods to eat without smoked ones.

PART III

DRINKING WATER

6

Water from Your Tap

Your morning cup of coffee contains the metals dissolved
from pipes overnight, unless you flush them out.
—HENRY A. SCHROEDER, M.D., 1974

Wherever you live, drinking water is an important element in
your daily life. The FDA considers water a food; water
should be healthful, not hazardous.

To some extent you do not control the quality of your
drinking water, except as you are willing to assume an active
role to improve local and federal regulations. But once water
reaches your home, you may want to consider various possi-
bilities for the control of the safety and quality of your drinking water. Items to be evaluated include water pipes (if you
are building or replumbing your home), water conditioners,
filters, purifiers, and distillers, or switching to bottled water.

Ideally, water pipes should be made of an inert substance that does not dissolve, vaporize, interact, or leach hazardous substances into water. Unfortunately, no type of pipe has yet been manufactured that meets all these qualifications. Some may be better than others, depending on the characteristics of your water. The pipe may affect the water, and, in turn, the water may affect the pipe. The first thing to determine is whether your area has soft or hard water. In general, soft water is acidic, while hard water with dissolved minerals is alkaline.

How can you tell if your water is soft or hard? Assuming that a water softener has not been installed, you can make a rough determination. If water makes suds easily with soap or detergent, consider it soft. If it forms laundry curds, it is hard. A bathtub ring reflects some degree of water hardness.

A simple home test for water hardness is to place a tablespoon of cold water on a small transparent glass dish. Put the glass on a source of very low heat, such as a radiator, and allow the water to evaporate. The water is relatively soft if the glass has only a very thin, spotty covering. The water is relatively hard if it has a small amount of clearly marked white residue; the more residue, the harder the water.

If you contact your local water-supply company or municipal water plant for specific information about the degree of hardness of your tap water, you may be given a specific figure of grains per gallon or parts per million, the official measure for water hardness. To understand the grains of hardness, you can perform a simple home test with tincture of green soap, available at a drugstore. Fill a small, clear glass bottle with an ounce of water. Using an eye dropper, add one drop of tincture of green soap, cover the bottle, shake it vigorously, and then time it. If the foam stands up for two to five minutes without breaking, the water is soft. If it breaks down, continue adding one additional drop for each test until the foam holds. The number of drops needed to keep the foam from breaking indicates the number of grains of mineral salts per gallon of water. For example, if the water requires five drops of tincture of green soap to retain the foam, the water is hard (17.1 ppm × 5 + 85.5 ppm per U.S. gallon, or 5 grain hardness).

If you have your water tested, you will be given its pH rating. The acidity-alkalinity of water is measured on a pH scale: 7.0 is neutral; lower numbers are acid; higher numbers are alkaline. Any water with a pH under 8.0 *can* be corrosive; under 7.0 *always* corrosive.

Soft water, especially if acidic, corrodes water pipes and can release toxic metals such as cadmium, lead, and cobalt. Hard water, generally alkaline, does not corrode pipes, but presents other problems. Calcium and magnesium salts in hard water precipitate calcium and magnesium bicarbonates on metal pipes. This gradual accretion forms a hard coating but actually *prevents* corrosion (except in rare instances when hard water is atypically acidic). So, if your water is hard, it helps prevent the release of toxic metals into your drinking water. Hard water contains many beneficial minerals. For both reasons, hard water is considered to be a positive factor in human health, while soft water is a negative one. Many studies support this idea.

Which type of water pipe suits your needs best? Obviously, there is no simple answer. If you live in a soft water area, you may have corroded pipes that leach toxic metals; in a hard water area, mineral accretions and coated pipes. Pipes made of lead, galvanized iron or steel, and copper will all leach some of their metals into your tap water if you live in a soft water area. Dissolved iron will stain your sink brown; copper, blue. Plastic piping may leach plastic as well as metals such as lead and cadmium into water. However, there are a few sensible guidelines you can follow, regardless of the type of water or pipes you have:

> Every morning flush out the water that has been standing in the pipes overnight. Do this before you draw water for drinking or cooking use.

If you bathe in the tub or shower before your kitchen preparations in the morning, you have flushed out the standing water in most of the piping, except for the length of pipe hooked to your kitchen tap.

If you have been away from home for an extended time, such as a vacation, allow more water to flush out.

Toxic metals in water are usually found in greater amounts when the water is hot, rather than cold. Use the cold water tap for drinking and food preparation.

Avoid boiling any drinking water for a long time. As water evaporates, toxic metals which may be present will concentrate.

If your water is naturally soft, set the hot water heater at the lowest level you find satisfactory. Hot water is far more corrosive than cold water to the metallic parts of water heaters, pipes, and plumbing accessories.

If you have pangs of guilt about wasting water by flushing the pipes in the morning, catch it in basins for use in hand dishwashing or laundering, watering plants, or cleaning floors.

7

Household Water Softeners

Till taught by pain,
Men really know not what good water's worth.
—LORD BYRON
Don Juan

Water can be so hard, however, that the disadvantages far outweigh the advantages. If you live in an area of extremely hard water, you know that it does not lather nor clean readily, and it may require more soap or detergent than soft water. You may be dismayed by dingy laundry. If you are a fastidious homemaker, streaks and stains on glassware may plague you. You struggle with personal grooming in bathing and shampooing. Along with all these nuisances, if you are a homeowner, you may be concerned with the fact that hard water lowers the efficiency of water heaters and boilers, and builds up scale inside of plumbing pipes, so maintenance is difficult and costly. If you experience these problems you may wish to install a water softener.

Two different types of water softeners depend on ionic exchange systems. The calcium and magnesium compounds in hard water, largely responsible for its inconveniences, are replaced by a sodium compound. Washing soda (sodium carbonate) is used in one system, by reacting with the calcium and magnesium ions and precipitating insoluble calcium and magnesium carbonates. The sodium ions remain in the softened water.

The other type of water softener uses zeolite, an insoluble granular silicate compound, either from natural sources (greensand or glauconite) or made synthetically (sodium aluminum silicate). Synthetic zeolite removes at least twice as many grains of hardness as natural zeolite.

One word of caution before you consider installation. Although a water softener offers you a means of overcoming the inconveniences of hard water, you should *not* use artificially softened water for drinking or cooking. *The harder your water is before treatment, and the higher the level of calcium and magnesium compounds it contains, the higher will be the sodium level found in the treated water.* So it is imperative that you have the plumber attach the water softener *only* to the hot-water system intended for laundry, dishwashing, and bathing, but *not* attached to any cold-water drinking taps. Sometimes, through carelessness or ignorance, water softeners have been improperly attached. Or persons unaware of the hazard habitually draw hot softened water for drinking and cooking purposes.

Four to five grains per gallon (70 to 80 ppm), which is moderate hardness, is considered optimum (see page 67). At this level, enough mineral content remains in the water to protect the piping and other conduits, but the water is not so hard as to seriously interfere with washing and cleansing.

Water of zero hardness is not only undesirable for your health, but would make rinsing soap and detergent from skin or utensils very difficult, requiring repeated rinsings to get rid of residues.

Remember that when you soften water, you increase the likelihood of corroding water pipes, since the calcium and magnesium ions removed from hard water are the very compounds which, if allowed to remain, would form scale and afford some protection to metal piping against corrosion. Remember also that, as said before, softened water should not be used for drinking or cooking. Nor should it be used to fill an aquarium, steam iron, or automobile battery. Do not have softened water installed for outside taps where you may use water to sprinkle lawns, shrubs, flower beds, or vegetable gardens, or for car and window washing.

Be aware that a water softener you install may adversely affect your neighbors, or theirs may affect you. In backflushing, large amounts of salt are washed into the drainage system, thereby increasing the sodium content of the water supply for households downstream. This might be a special hazard for people on low-sodium diets.

Basically there are four types of zeolite water-softening devices for household use. The terms designate the method of regenerating the zeolite: fully automatic, semiautomatic, automatic, and manual.

For a listing of brand names, names of companies that market water softeners for household use, and advisory services, see *Consumer Bulletin*, published by Consumers' Research, Inc., November 1965. For more recent additions, consult the Yellow Pages of the telephone directory. Also, contact your Chamber of Commerce for a list of local suppliers, and check their reputations with the Better Business Bureau or your consumer-protection agency.

Look for water softeners that bear a "Water Conditioning Foundation Gold Seal." This emblem is granted to water-conditioning equipment which has been tested and validated by the Water Conditioning Foundation under standards approved by the Federal Housing Administration.

8

Household Water Filters, Purifiers, and Distillers

When water chokes you,
what are you to drink to wash it down?
—ARISTOTLE

If you are in an area where the drinking water is unsafe or unpalatable, you may be interested in household devices intended to improve it. Such devices flood the marketplace. At times the promotional literature and sales pitches make claims that are exaggerated, misleading, or patently false. It is important to separate fact from fallacy. What can these devices do for you? What are their limitations and possible hazards?

Water filters, the commonest and least expensive devices to improve water, are able to remove certain undesirable substances to some extent. Some filters depend on activated charcoal, and, at times, for greater effectiveness, the charcoal is combined with cellulose filters, sand, and pebbles.

Some activated charcoal filters can be installed permanently to your kitchen tap, to your household water main, or to your water supply before it reaches a tap, as in a pipe beneath a sink. Another type is a unit that rests on a kitchen counter near your sink, and is connected by tubing to a faucet. Other devices, intended for use in traveling, are portable.

Some filter systems are described as being capable of removing unwanted contaminants, including minerals, toxic metals, rust, scale, objectionable odors and tastes, discolorations, turgidity, algae, and bacteria. Some filters are advertised as being able to produce crystal-clear, safe drinking water. Regard such statements as puffery. Even well-equipped water-supply systems, using the most sophisticated techniques, can't accomplish all these feats, and household systems are far simpler.

What can you expect from a household filter? *To some degree*, it can filter out turbidity and produce cleaner-looking water. *To some* extent, it can reduce chlorine, rust, dirt, objectionable odors and tastes, and suspended matter.

What *can't* you expect? The filter probably cannot remove fluoride, toxic metals, or organic chemical contaminants. Nor can it eliminate completely objectionable odors, tastes, discolorations, and turbidity. If these features are present, even at such exceedingly low levels as one part per billion (1 ppb), they may still be detectable.

Don't rely on a filter to remove bacteria from your drinking water. On the contrary, such false assurance can be dangerous. Activated charcoal can absorb on its surface lots of chlorine, present in the water from the municipal-water-supply treatment. If the filter removes the small amount of residual chlorine from the tap water, the filtered water loses its ability to combat the bacteria which are then free to multiply. The filter, itself, can become a breeding ground where bacteria may thrive.

How long will your filter be effective? You have no way of knowing precisely whether a household filter is still functioning properly after some use, or if it has become ineffective. The higher the level of water contamination, the shorter will be the filter's usefulness, and the more frequently you will need to replace it.

The Environmental Protection Agency conducted tests on the effectiveness of some home water filters, and the results were not very encouraging. The agency found that many of the filters do not remove enough of the harmful chemical load in water to make their use worthwhile. One exception was the Everpure Model QC4-THM filter, which removed 93 percent of trihalomethanes (THMs), which are carcinogenic compounds that include chloroform. This device uses two filter cartridges, one of granular carbon and the other of powdered carbon.

The Environmental Defense Fund has suggested what it considers an inexpensive and safer homemade alternative to home filter units. You will need a large funnel, a large clean jar, coffee filter papers, and washed granulated activated carbon. You can purchase the carbon in one-pound bags from Walnut Acres, Penns Creek, Pa. 17862. To wash the carbon, place it in a jar, fill the jar with water, cover, and shake it. Allow the carbon to settle, and then pour off the water. Repeat this cleansing operation until the poured off water is entirely clear.

Place a coffee filter paper in the funnel, and set the funnel so that it rests at the top of a large jar. Fill it one-quarter full with washed carbon. Slowly pour tap water through the funnel. The water that filters through the jar should be stored in the refrigerator until it is used. Change the carbon every three weeks, or after you have filtered twenty gallons of water.

A household water purifier should make your water safe by killing bacteria.

Some purifiers depend on ultraviolet light for their germicidal effectiveness. Whether or not this type of purifier works well depends on many factors: the wave length, the intensity of ultraviolet energy, the rate of water flowing through the purifier, the water's temperature (the purifier's effectiveness is decreased below about 50° F.), and the presence of color, turbidity, iron, or organic substances in the water.

A flow control device on an ultraviolet purifier limits the water flow to about eight gallons a minute to ensure the germicidal effectiveness of the ultraviolet light. Otherwise the device might be ineffective during peak periods of water use.

One type of water purifier, known as the "silver oligdynamic method," (oligdynamic: active in small amounts) makes use of minute amounts of silver ions. Silver ions are released from a carrier, such as activated charcoal or diatomaceous earth, as the water passes through the filter. Any pathogenic bacteria present ingest the silver and are killed by it. Some silver passes through the filter, enters the treated water, and prevents new bacterial growth. The remaining silver ions in the treated water may exert some nontoxic biologic effects in humans, but they differ from toxic silver salts.

Another water-purifying system, known as "reverse osmosis," makes use of a semipermeable membrane to remove some impurities. The more pressure at the faucet, the higher the degree of purification with this system. But reverse osmosis will not necessarily provide uniform water quality. The membrane does not prevent certain contaminants such as iron and nitrate molecules from passing through.

You can purify water and kill any bacteria it may contain simply by boiling it. Although boiling will remove some volatile organic chemicals (such as chlorine), it will concentrate others (such as fluorides).

A household distiller is the most efficient and thorough way to purify water. Distilling water is a slow process that forms rain and snow. Using the same basic principles but controlling the heat source, a household distiller purifies water by vaporizing and then condensing it. Most, but not all, of the contaminants remain behind. But distilling is not especially effective in removing carcinogens. For example, benzene forms a "constant boiling mixture" with water and will be carried over with the distillate.

For the highest chemical purity, water must be distilled several times. Then it should be completely free of all foreign matter, dirt, rust, detergents, pesticides, and heavy metals. But it will also be free of many major and minor minerals essential to good health. Distilled water has limited use under special circumstances. It may be used therapeutically, or for emergencies when it is the only temporary, practical supply of drinking water. For long-term use, however, distilled water is undesirable as a source of drinking water.

Some people favor the use of distilled water for all drinking and cooking purposes. They contend that the inorganic minerals in drinking water are not utilized by the body, are deposited, and result in health problems. This viewpoint is contrary to fact. Both inorganic minerals (from earth or sea salts) and organic minerals (from foods) are necessary to life and used in the body.

Water in which minerals are dissolved contains electrolytes, in chemical equilibrium with the minerals before they went into solution. When they are in a state of balance, the water is *isotonic*. Distilled water, devoid of minerals, is *hypotonic*; sea water, loaded with minerals, is *hypertonic*. Fluids in the human body include blood, urine, lymphs, tears, sweat, and saliva and are found in the extracellular and intracellular systems and in the cerebrospinal system. All these fluids exist in a hypertonic state. Hypertonic fluids are secreted or excreted through organs such as the kidneys, liver, skin, and lungs.

If you drink distilled water, which lacks minerals, the water will be absorbed quickly and directly into your bloodstream. In the process, minerals are extracted from your body membranes, are absorbed by the water, and are excreted from your body. So, *distilled water, by not contributing any minerals to your body, upsets the electrolyte equilibrium and depletes your body of its mineral supply.* If you continue to drink distilled water over a period of time, and the electrolyte system remains hypotonic, you will develop mineral deficiencies and health problems. Some people who drink distilled water use supplements such as sea water, kelp, or mineral tablets. Although such supplements will help to provide some minerals, there is no assurance that they achieve and maintain an ideal state of isotonic balance.

Remember that household filters and purifiers are unreliable for water that may be contaminated by bacteria or other pathogens. In an emergency, boil your drinking water for at least twenty minutes. For a listing of household filters and purifiers, and a description of specific devices, see *Consumer Bulletin* (Consumers' Research, Inc.), January 1973. The performance characteristics of water filters are rated for effectiveness in removing iron, copper, and odors, and their ability to reduce turbidity, sudsing, suspended matter, and so forth. For additional devices, consult the Yellow Pages of the telephone directory. Also contact your Chamber of Commerce for a list of local suppliers, and double-check with the Better Business Bureau or your consumer-protection agency regarding their reputation.

If you want to know the composition of your drinking water, you can arrange to have it tested by a state agency. You will be given directions for preparing and shipping the sample. Usually the charge is nominal. Or, you can have a more extensive test made for $25 by sending a check or money order to Water Test, Soil and Health Foundation, 33 East Minor Street, Emmaus, Pa. 18049. The analysis of your household water is compared to standards established by the EPA, and measures calcium, magnesium, potassium, sodium, arsenic, barium, cadmium, cobalt, copper, iron, lead, manganese, mercury, selenium, silver, zinc, and chromium.

If your water is unsafe or unpalatable, you may wish to seek another source, such as natural spring water or well water. What is the difference? Spring water flows out of the earth without being pumped; well water must be pumped to the earth's surface.

Or, lacking any nearby natural spring or well water, you can switch to bottled water. You have a large selection, both domestic and imported, bubbly or still, depending on your taste and budget. Many reputable companies list the mineral contents of their water or, upon request, will supply them. The sodium content is of special interest if you are on a low-sodium diet. Not all mineral waters are low in sodium.

Unfortunately, no mandatory nationwide standards exist for the chemical and bacteriological quality of bottled waters. The American Bottled Water Association (ABWA), representing most of the industry, follows federal drinking-water standards and good sanitation practices. ABWA attempts to police itself, and withdraws its certification of noncomplying members. The EPA samplings of bottled water in 1973 revealed that some water was dirty, bacterially contaminated, and contained excessive levels of toxic substances such as lead. Deficiencies were found in all the facilities surveyed. The EPA recommended that water-bottling plants be inspected regularly and that quality control be improved.

Read labels carefully. Don't be misled by the word "spring," as in "spring fresh" and "spring pure," for such terms do *not* mean natural spring water. About 75 percent of bottled water sold in the U.S. is not natural mineral water but reconstituted or "formulated" from local tap water by distillation, filtration, deionization, or electrodialysis, and/or supplemented with calcium, sodium, magnesium, or carbonates.

Make certain that the bottle cap is well sealed and that no substitution is made of its contents. Choose glass containers for bottled water, which will not transmit odor, taste, or bacteria to it.

Principal Sources

Pots and Pans

Accum, Frederick. *Death in the Pot*. London, circa 1830.

Archives of Industrial Health, May 1957.

British Medical Journal, March 26, 1932; April 16, 1932; May 7, 1932.

Chemical and Engineering News, June 27, 1977; August 29, 1977; October 17, 1977; November 21, 1977; February 6, 1978.

Chemistry, July/August 1971; May 1972.

Consumer Bulletin, July 1961; May 1965; May 1966; February 1967; April 1972.

Consumer Reports, March 1962; January 1966; October 1967; March 1968; October 1971.

Consumers' Research, December 1976; February 1978.

FDA Consumer, July/August 1977; September 1977.

Food & Cosmetics Toxicology, August 1965; September 1965; June 1968.

Journal of the American Dietetics Association, vol. 60, 1972.

Medical Tribune, January 12, 1972.

Medical World News, April 16, 1971.

Monier-Williams, G. W. *Trace Elements in Food*. New York: Wiley, 1950.

New York Times, July 9, 1977; July 18, 1977.

Nutrition Today, March/April 1972.

Safety of Cooking Utensils, FDA Consumer Memo, July 1971.

Schroeder, Henry A. *The Poisons Around Us*. Bloomington, Ind.: Indiana University Press, 1974.

————. *Pollution, Profits & Progress*. Brattleboro, Vt.: Stephen Greene, 1971.

————. *The Trace Elements & Man*. Old Greenwich, Conn.: Devin-Adair, 1973.

Science News, September 17, 1977.

Toxic Metals in Dinnerware, FDA Fact Sheet, July 1971.

Underwood, E. J. *Trace Elements in Human & Animal Nutrition*. 3rd ed. New York: Academic Press, 1971.

Utah Science, September 1972.

Wall Street Journal, March 30, 1962.

Kitchenware

Chemical and Engineering News, August 29, 1977; October 17, 1977; November 21, 1977; January 16, 1978; February 6, 1978; January 1, 1979.

Chemistry, September 1971.

Consumers' Research, November 1977.

FDA Consumer, July/August 1977; September 1977; June 1978; March 1979.

Good Housekeeping, March 1976.

HEW News Release, January 31, 1978; December 15, 1978.

New York Times, July 9, 1977; July 18, 1977; February 1, 1978.

Microwave Ovens

Brodeur, Paul. *The Zapping of America.* New York: W. W. Norton, 1977.
Co-Op News, Hanover, N. H., November/December 1979.
Environment, May 1970; June 1974.
FDA Enforcement Report, January 25, 1977; February 23, 1977; September 14, 1977.
Hearings on Microwaves, Committee on Commerce, U.S. Senate, May 1968.
HEW News Release, June 25, 1975; January 11, 1977.
Journal of Applied Physiology, July 1962; December 1967.
Life Sciences, vol. 7, part II, 1968.
Medical World News, August 2, 1974.
More Protection from Microwave Hazards Needed, GAO Report, November 30, 1978.
New York Times, August 9, 1977; September 15, 1977; December 27, 1977.

Charcoal Grills

Chemical and Engineering News, May 22, 1978.
Fast Service, June 1978; August 1978.
Food & Cosmetics Toxicology, vol. 5, 1967.
Industrial Medicine & Surgery, February 1965.
New York Times, December 26, 1977; May 2, 1978.
Science, February 28, 1975.
Science News, May 20, 1978.

Food-smoking Chambers

Archives of Pathology, April 1961.
Chemical and Engineering News, January 2, 1978.
Food & Cosmetics Toxicology, February 1966, August 1972.

Hospital Practice, October 1975.
Journal of Agricultural and Food Chemistry, vol. 14, 1966; vol. 24, 1976.
Journal of the National Cancer Institute, February 1960.
Newsweek, December 4, 1961.

Water from Your Tap

Environmental Science and Technology, December 1974.
Journal of Occupational Medicine, February 1965.
Medical Tribune, August 23, 1978.
Medical World News, October 12, 1973.
New York Times, January 3, 1977.
Science News, January 7, 1978.
Wall Street Journal, December 23, 1977.
Water, the Yearbook of Agriculture, USDA. Washington, D.C.: U.S. Government Printing Office, 1955.

Household Water Softeners

Consumer Bulletin, February 1960; November 1966.
Consumers' Research, September 1974.
Journal of the American Dietetics Association, vol. 25, 1949.
Journal of the American Medical Association, January 31, 1951; August 8, 1953; May 20, 1961.

Household Water Filters, Purifiers, and Distillers

Consumer Bulletin, October 1965; September 1970; January 1973.
Consumer Reports, April 1972; July 1972; June 1974; July 1974; August 1974.
Food and Drug Packaging, March 5, 1975.
Horizon, August 1978.
Medical Tribune, September 13, 1972.
New York Times, March 2, 1973; January 12, 1977; June 26, 1977; November 14, 1979.
Wall Street Journal, January 10, 1973.

Index

acidic water, 66–67
activated charcoal, 73
agateware, 4
alkaline water, 66–67
aluminum: and Alzheimer's disease,
    8; and consumer goods, 9;
    foil, 44–45; in food, 5–9; from
    industrial pollution, 8; in
    municipal drinking water, 9; in
    phosphate compound, 9
aluminumware, 4–10; bacterial
    growth in, 15; the FDA and, 5
Alzheimer's disease, 8
American Bottled Water Association
    (ABWA), 78
anodized aluminumware, 15
antimony: oxide compound in
    enamelware, 15; in pewter, 38
antioxidants in food wrappers, 43
Aristotle, 72

baked enamel cookware, 10, 28
Bartelmes, 28
beef extract, formation of mutagens
    in cooked, 59
benzo(a)pyrene, 57, 61
boil-in-bag pouches, 24–26

bottled water, 78–79
Byron, 69

cadmium: in decorated glassware,
    36; in food and beverage
    processing, 40; in food contacts,
    39–40; in glazed earthenware,
    11–12; in ice trays, 39; in metalware,
    39–40; in plastic food wraps, 42; in
    plastic water pipes, 67; in serving
    trays, 39; in water, 67
calcium bicarbonate, 67
Cassidy, Lorne, 8
cast ironware, 10
cellophane, 43–44
cellulose in water filters, 73
ceramic-coated cookware, 10, 27
charcoal grills, 57–59
Childhood Lead-Poison Prevention
    Program, 35
chloroform, 74
chromium in stainless steel, 29–30
clay earthenware, 10–12; cadmium
    in, 11–12; glazed, 10–12; lead in,
    11–12; unglazed, 10–12
cobalt in water, 67
columbium in stainless steel, 30
Commoner, Barry, 59

Consumer Bulletin: on water filters, 72–75; on water purifiers, 75; on water softeners, 71
Consumer Product Safety Commission (CPSC), 36
Consumers Union, statement about microwave ovens by, 54–55
copper: arsenite, 13; poisoning from, 13–14; sensitivity to, 31; sulfate, 13; and Wilson's disease, 14
copperware, 12–14
Corning portable counter saver, 47
cutting boards, food, 46–47

decals and lead, 33–37
decorated glassware, 33–37
dinnerware, 37–39
distilled water, 76–77
drinking water, 65–79
Du Pont: cellophane by, 43; nonstick cookware by, 19–21; nonstick fluorocarbon finishes by, 23

earthenware, 14
electrolytes, 76–77
enamelware, 14–16
Environmental Protection Agency (EPA): and water filters, 74; and water safety standards, 78
Everpure water filter, 74

filter unit, home, 74–75
flameproof glass cookware, 16
fluoride in cooking utensils, 21
fluorocarbon-coated nonstick cookware, 16
fluorocarbon resins, trade names of, 23
food cutting boards, 46–47
Food and Drug Administration (FDA): on aluminum food contacts, 4; Bureau of Radiological Health (BRH) of the, 54–55; on clay earthenware, 11; on decorated glassware, 34; on formaldehyde resin, 38; on micro-

wave ovens, 54–56; on nonstick cookware, 20–21; on plastic cooking bags, 25; on water, 65
food-smoking chambers, 60–61
food storage wrappers, 43–46
formaldehyde in early plastics, 42

galvanized zinc, 40–42
glass: cookware of, 16–17; food storage containers of, 41–42
grains of water hardness, definition of, 66–67
graniteware, 17
grills, charcoal, 57–59

hamburgers, formation of mutagens in cooked, 59
hard water, 66–67
heat tint in stainless steel, 31
hemochromatosis, 18
hexamethyleneamine, 25
hypertonic water, 76
hypoallergenic cellophane, 44
hypotonic water, 76

ionic exchange system, 70
iron: absorption from cookware, 18; in water, 67
ironware, 17–19; seasoning of, 18

Kettering Laboratory, and aluminum, 7
kitchenware, 33–47

labeling of bottled water, 78–79
lead: in decals, 33–37; in earthen dinnerware, 37–38; in glazed earthenware, 11–12; in metalware, 39; in pewter dinnerware, 38; in plastic pipes, 67; in porcelain dinnerware, 37–38; in water, 67
Lijinsky, William, 60

magnesium bicarbonate, 67
melamine formaldehyde resin, 38

Mellon Institute of Industrial Research, and aluminum, 6–7
metal foils as food wraps, 44–45
metalware, 39–42
methylamine, 25
microwave ovens, 51–56; browning plates in, 59; defective viewports in, 54; proposed safety standards for, 54; radiation leakage from, 53–56; safety interlock failures in, 54–56; U.S. Surgeon General's Ad Hoc Task Force Report on, 54
Modeno, Franco Marzulli, 15
molybdenum in stainless steel, 30
Monier-Williams, G. W.: on aluminum safety, 7–8; on metal poisoning, 3
Moore, Carl V., 18

n-hexylamine, 25
nickel: sensitivity to, 31; in stainless steel, 30
nonstick cookware, 19–24; bacterial growth in, 15; resin finishes of, 21–23
nylon films, degradation of, 25

opaque ceramic glass cookware, 16–17, 24
ovenproof glass cookware, 16

parchment paper, 26–27
patapar paper, 44
Percival, Thomas, 12
pesticide-treated shelf liner, 43
Peterson, Esther, 26
pewter dinnerware, 38
Pfeiffer, Carl C., 8
plastic: acrylic coating, 28; cooking bags, 24–27; dinnerware, 38; food storage containers, 42; food wrappers, 45–46; plasticizers in, 24; water piping, 67
polyamide coating, 28
polyamide-imide plastic resins, 22
polybutene, 26

polycyclic aromatic hydrocarbons (PAH), 57–59
polyester: low-density film of, 26; medium-density web of, 26
polyethylene as food wrap, 45–46
polyphenylene sulfide plastic resins, 22
polyvinyl chloride (PVC) in plastic food wraps, 45–46
porcelain-enamel coated cookware, 27–29
porcelainized cookware, 29
porcelain-on-steel cookware, 27
potable water, 65–68
pots and pans, 1–32
Pott, Percival, 58
purifiers, water, 75

Randolph, Theron G., 33
resin finishes of nonstick cookware, 21–23, 29
reverse osmosis system, 75
Reynolds Metal Company, 25

Schroeder, Henry A.: on aluminum safety, 7–8; on water pipes, 65
Shubik, Philippe, 60
silicone-coated nonstick cookware, 22–23, 29
silver oligodynamic system, 75
silver-plated metalware, 41–42
smoked food chambers, 60–61
smoked food and PAH, 60–61
soapstone griddle, 24
soft water, 66–67
soy lecithin liquid, 24
Spittler, Thomas M., 35–36
spring water, 78–79
stainless steel cookware, 29–32
stoneware, 32
storage containers for food, 42–43
storage wrappers for food, 43–46
Susskind, Charles, 51

tapwater, 65–68
Teflon cookware, 19–24, 32

tests for water hardness, 66–67
tetrafluorethylene (TFE), 19–21
tin oxide and enamelware, 15
tinware, 32
titanium in stainless steel, 30
trihalomethanes (THMs), 74

ultraviolet light, 75
Underwood, E. J., 7–8
Upton, Arthur C., 57–58
utensils, cooking, 1–32

Vita Wrap, 27
vitreous enamel cookware, 27, 32

water: bottled, 78–79; distillers, 76–77; drinking, 65–79; filters, 72–75; labeling of bottled, 78–79; pipes for, 65–68; purifiers, 75; safety tests for drinking, 77–78; softeners, 69–71; spring, 78–79; well, 78
Wilson's disease, 14
wooden dinnerware, 38
woodenware, 46–47
World Health Organization (WHO): monitoring toxic metals by, 12

zeolite, 70
zinc: antimony in, 40–41; arsenic in, 40; cadmium in, 40; galvanizing with, 40–42; iron in, 40–41; metalware and, 40–41